# THE COOL PURPOSE

*The Journey of Danfoss Turbocor in Engineering a Better Future*

Ricardo Schneider

Renown
PUBLISHING

**Renown Publishing**
**www.renownpublishing.com**

**The Cool Purpose / Ricardo Schneider**
Paperback ISBN-13: 978-1-960236-40-1
Hardback ISBN-13: 978-1-960236-41-8

# CONTENTS

# Foreword by Jørgen M. Clausen

When my father, Mads Clausen, began shaping small metal parts in the barn behind our home on the Danish island of Als and later pioneered the variable speed drive, he was not thinking about the global impact his legacy would have a hundred years later. He simply believed that electronics and mechanics, brought together with purpose, could solve meaningful problems.

That belief became the foundation of Danfoss. It is part of our DNA to innovate with purpose, stay close to customers and think long-term.

Nearly a century later, I have seen how that strong DNA can cross continents and cultures. The journey of Danfoss Turbocor is one of those examples where two very different worlds meet and create something entirely new that redefines an industry.

In Australia, Ron Conry and his small team challenged a centrifugal compressor industry that had changed little since the days of Willis Carrier. They imagined an oil-free, frictionless, variable-speed compressor unconstrained by traditional design. Their courage and refusal to accept "that's how it's always been done" reminded me of the early days of Danfoss when my father founded Danfoss.

When I first saw Turbocor at the AHR Expo in Chicago in 2003, I remember being struck by how small the centrifugal compressor was compared to traditional ones, but even more by the seamless integration of power electronics and magnetic bearing technology. In that moment, I felt I was looking at the future. It made me absolutely determined that this was what we had to do, whatever the cost!

We decided to acquire the company at a time when the technology was far from developed. It required a significant investment, and we did not know whether we would succeed. Many of our large competitors examined the same opportunity, but they chose not to move. Honestly, it could just as easily have become a costly mistake.

Bringing Turbocor into the Danfoss family was not easy. Transformations are never easy but it was the right decision. Together, our teams built something neither could have achieved alone. The team who built Danfoss Turbocor lived our DNA, and they still do. The story of Turbocor is simply a story about innovation, courage, strong performance, entrepreneurship, and the belief that innovation and technology should improve people's lives and engineer a better future. It also confirms that technological leadership is built over decades and not financial quarters.

This spirit lived not only in the early pioneers, but also in the leaders who transformed Danfoss Turbocor from a fragile startup into a global force, crafting a winning strategy, deepening customer relationships, and building the foundation for Turbocor's long-term success.

Ricardo Schneider, supported by Kim Fausing, led Turbocor with that same founder's mindset that my father had, balancing bold innovation with execution, nurturing a culture of ownership, and guiding a breakthrough technology into a true industry standard.

I smiled when I read about Turbocor traveling across the United States in a trailer, introduced to customers one by one. Great technologies are rarely developed in boardrooms. They begin with belief, dedication and a clear customer focus.

Today, Danfoss Turbocor stands as proof of what happens when vision meets perseverance and when people from different backgrounds choose to trust one another. What began as a bold engineering idea has become a central part of our broad portfolio and plays an important role in areas such as data center cooling, where efficiency and reliability are critical.

It is a story of innovation, courage, and the quiet determination of individuals who believed they could make a difference.

As you read this book, I hope you feel the same pride I do: pride in the pioneers, from Gorrie and Carrier to Mads and Ron; pride in the Danfoss Turbocor team, who turned ideas into reality; and pride in the oil-free community that Danfoss

Turbocor and Danfoss have had the privilege to embody in our culture of engineering a better future.

INTRODUCTION

# A Vision Takes Shape

From the moment we wake up in climate-controlled homes to the food we consume, the hospitals we rely on, and the data centers powering our digital footprint, our lives are supported by an industry that quietly enables modern civilization.

Heating, Ventilation, Air Conditioning, and Refrigeration (HVAC-R) is a massive global sector valued at over a trillion dollars, with steady growth driven by urbanization, climate change, and technological innovation. In the U. S., Europe, and Asia, it keeps people comfortable and their food from spoiling. It allows vaccines to be safely stored and industrial manufacturing to create sustainable building design for cooling and heating. Despite being taken for granted, this infrastructure is indispensable. Without it, daily life and technological progress would come to a standstill.

As artificial intelligence continues to expand, the demand for cooling in data centers has surged dramatically. AI workloads generate intense heat, requiring advanced HVAC-R systems to maintain safe operating conditions. Globally, HVAC-R systems consume over one-third of building energy. This energy footprint contributes significantly to climate change, especially when powered by fossil fuels and using high global warming potential (GWP) refrigerants.

To address this, the industry must innovate, shifting toward higher energy efficiency, renewable energy integration, and climate-friendly refrigerants, aiming to balance the growing demand for comfort and computing with environmental sustainability.

Thankfully, a group of visionary engineers on opposite sides of the world have created a company that is well on its way to solving that problem.

Ninety years ago, the son of Danish farmers introduced a mechanical expansion valve to improve energy efficiency and then, twenty years later, massive production of compressors for refrigerators. Sixty years later, an Australian entrepreneur charted a breakthrough that would eliminate oil usage from compressors for refrigeration and air conditioning.

The paths of their teams would collide at the dawn of the twenty-first century and usher in a new era at the moment the world was clamoring for newer, faster, cleaner cooling infrastructure. But to understand how this came to be, we have to step back to the days of the horse and buggy, before even trains and telegraphs were connecting the world—back to when Dr. John Gorrie had a problem with the heat.

Specifically, he found himself battling the hot, sticky climate of Apalachicola, Florida, a small city about seventy-six miles from Tallahassee where the Apalachicola River met the Gulf. In the 1840s, it was a bustling cotton port, but come summer, the town transformed into something else entirely. Diseases like yellow fever and malaria had a death grip on the populace. Gorrie, who headed two hospitals, set about researching what caused these ailments that swept through the territory and other cities along the East Coast every summer.

He was, at the time, among other researchers who bought into the theory of bad air—or "mala aria," as the Italians termed it. This gave rise to the modern disease appellation of "malaria," which we now know is spread via mosquitoes. He was, of course, mistaken about the cause of those diseases, but as he focused on the air, he decided that the best solution would be to make it cooler.

Gorrie put blocks of ice into basins and hung them from the ceiling in treatment rooms, allowing the cold air to settle onto the patients. The ice supply in Florida was, understandably, difficult to come by. Blocks had to be cut from lakes in Northeastern states, packed in sawdust, and shipped down the coast.

His eureka moment involved using a compressor to compress air, a heat exchanger to cool that air, and an expansion device to expand it, thus dropping the temperature enough to create ice. These were the same principles that lasted, with modification, until today, used in refrigeration and air conditioning for food preservation or cooling data centers.

His labor of passion eventually resulted in U. S. Patent 8,080, for the "Improved process for the artificial production of ice." By the time the 1850s arrived, he was unfortunately unable to get the funding he needed. He faced opposition and slander from the ice industry in the North, which was afraid he would disrupt their profitable business model, and from religious zealots who felt it was against the order of creation to make ice.

The next seventy years saw refrigeration technology go through several iterations, most notably in the form of Australian James Harrison's creations, which used Gorrie's principles in a closed circuit system, using ether as refrigerant fluid for the purpose of food preservation. He received patents in Australia in 1855 and the United Kingdom in 1856. In 1873, the Australian Institute of Refrigeration, Air Conditioning, and Heating gave him the gold medal at their prestigious Melbourne Exhibition.

But it wasn't until Willis Carrier invented the first centrifugal compressor in 1922 that air conditioning design was truly revolutionized, resulting in the ability to cool larger spaces more effectively. This meant that air conditioning was no longer used primarily for industrial processes, as had been slowly developing, but also for improving human comfort in various environments.

Ten years later, a Danish engineer and entrepreneur by the name of Mads Clausen had an idea that would lead to the next leap in cooling technology. Mads was the son of a farmer who grew up in the rural community of Als, Denmark, where flat farmland stretched to the horizon and North Sea winds bent the sparse trees permanently eastward. This was practical country, where nothing was wasted and innovation came from necessity. Years later, in the barn behind his father's house, Mads would set up his first workshop, arranging his tools with the same precision his ancestors had used to organize their plowing schedules. He had opted to become an engineer instead of a farmer, seeing as how he would not inherit his father's land, but also because of his fascination with all things mechanical. Mads's grandfather had built pumps and spinning wheels in his workshop, and this spurred the younger Clausen to attain his mechanical engineering degree in 1927.

He signed on with Brødrene Gram (Gram Brothers), designing components for refrigeration systems. At the time, refrigerators consisted of four major components, according to Mads's son Jørgen Clausen: a compressor, an expansion valve, and two heat exchangers, plus whatever insulation was needed. It was logical to start with the expansion valve to form a business. The compressor would come later, when Mads set the first production in the 1953.

Why the expansion valve? In heating, ventilation, and air conditioning (HVAC-R) systems, the most critical and energy-intensive component is the compressor, which compresses vapor, thus lowering the temperature of either water or air passing through the chiller components, but the thermal expansion valve is the device that adjusts the refrigerant flow, acting as the built-in intelligence to make the system efficient and reliable.

During the 1920s and 1930s, as the world entered the throes of the Great Depression, a key component, the thermostatic expansion valve, was manufactured solely in the United States. There was a ban on purchasing those components, which meant to get his hands on that vital valve, Mads would have to make one himself.

He worked hard to hone the perfect valve, staying long after hours at Brødrene Gram, sometimes sleeping during the day because he was up all night, according to Jørgen.

Eventually, after being reprimanded, Mads resigned in late 1933. Within eight months, he established his own company for the production of those valves: Danfoss.

Jørgen Clausen was just a small boy, but throughout his childhood in the 1960s, he watched the company grow. When Mads made it home for dinner, the conversation inevitably turned to pressure controls and refrigerant cycles. At that point, Danfoss built purely mechanical components, but he would emphasize to Jørgen that "electronics are the next thing."

It was during the 1950s and 1960s that Danfoss really positioned itself as one of the leading suppliers of refrigeration technology, producing the fractional compressors for refrigerators and freezer. Between 1953 and 2010, the company churned out an astonishing 285 million compressors, mainly in Flensburg, Germany, on top of those produced by EMBRACO in Brazil following the establishment of a license agreement in 1971.

Throughout those heady years, Mads Clausen not only focused on expanding the business, but also on supporting his employees, the people who were building his company into something great. He formally initiated a Welfare and Interest Office in 1956 that contributed to many foundations and departments Danfoss had spun off to provide for its employees, including one that supported those who had battled illness as they got back on their feet. This was the Danish way: take care of your people; think in decades, not quarters; build something that will outlast you. It was a philosophy that would both strengthen and strain the company when it eventually met its Australian counterpart.

Mads, the entrepreneur with a vision, thrived on consensus and relied on Bitten, his wife, to aid in the decision-making process. She had a strong personality that she used to complement her husband's entrepreneurial vision and creative energy. Often they would have businesspeople visiting, but since they still lived in a rural area, there were no accommodations available. The Clausens responded by opening up their home and hosting dinners.

Toward the end of Mads's life, he positioned Danfoss as the first mass producer of a variable speed frequency converter, often called a Variable Frequency Drive (or VFD), allowing Danfoss to control the speed of a three-phase induction motor. This was key technology for managing process control, system optimization, and energy efficiency, and it was later applied to compressors, a long-standing Danfoss goal.

Mads had the foresight to purchase a license to manufacture hydraulic orbit motors from Char-Lynn, a Minneapolis, Minnesota, company in 1961. Danfoss had focused primarily on heating and refrigeration, but at one point, Mads's chief engineer came to him about the metallurgy involved.

"He said to my father, 'You know, now we are making compressors, and that is something built of cast iron,'" Jørgen remembered. "In the past, it was either copper or brass—soft materials—and the cast iron is much more difficult to work with. He asked my father, 'Why don't we make hydraulics?' And my father said, 'Yes, why not?'"

It turned out another Danfoss employee had a cousin who was a hydraulics engineer and had recently completed his military service. With help and the changes in materials, Danfoss eventually became the world's largest mobile hydraulics manufacturer.

When Mads passed away, he left a legacy not only in his business, but also in his family. "At the time, I had to decide what kind of engineer I wanted to be," Jørgen explained. "Of course, I listened to what my father had said, so I chose to become an electrical engineer."

Jørgen extended his training by traveling to the United States and earning an MBA from the University of Wisconsin. His first job upon returning to Europe was at Grundfos, a pump manufacturer in Denmark, where he worked in marketing.

"As a marketing manager, I had to get to know everything about pumps, especially water pumps, turbomachinery, and centrifugal pumps, but also speed regulation, [which] I knew from Danfoss," Jørgen said. "After two years, I left Grundfos, and then I worked at other places before coming to Danfoss in 1981 as a corporate research manager."

His in-the-field education continued as he studied frequency converters, necessary because of Danfoss's emphasis on refrigeration and cooling. This bolstered his understanding of the specialized technology his colleagues had been working on. Ten years later, he became manager for the company's hydraulic business, which led to about $250 million in revenue annually.

Jørgen's rise through the company continued until he was appointed CEO and president in 1996, something he had not envisioned at the start of his career. He threw himself wholeheartedly into running the business his father had founded sixty years before.

Change was coming to Danfoss, however, because the commoditized business of compressors was faltering. They were facing increased competition from EMBRACO-Brazil and newcomers from China, a market that was drawing manufacturing from around the world. In 1993, Danfoss acquired Maneurop, a company that was introducing a new technology for commercial air conditioning called scroll compression, which had far fewer moving parts and was less expensive than a screw compressor and piston compressor. It was popular in commercial air conditioning and was the perfect way for Danfoss to start moving away from the commodity business of compressors for refrigerators.

At the same time, there was another revolution taking place on the other side of the world, in Australia, where Ron Conry had a vision for cooling technology that was about to change the world: a compressor that required no oil, levitated in a magnetic field, and was fully integrated with variable speed and sensors.

Since John Gorrie, James Harrison, and Willis Carrier, all cooling system compressors required lubrication to reduce the friction between the moving parts and prolong the lifetime of the compressors. As the compressor pumped refrigerant gasses, it also pumped some of the oil that lubricated the drive shaft, requiring complex adjustments and considerable cost to bring the oil back to the compressor. In most applications, the compressor also heated the oil up, requiring oil coolers to keep the oil on its lubricant characteristics.

Additionally, most of the commercial compressors were still mechanical and ancient in terms of their designs. They lacked variable speed digital devices, for example, and the mature technologies that featured low energy efficiency were controlled by a few international companies.

Then there was the size. Centrifugal compressors were massive machines that weighed anywhere from 500 to 5,000 kilograms. The loud noise and vibration they produced relegated them to the far corners of sizeable buildings, and in cases where that wasn't possible, soundproofing materials had to be installed to remediate the racket.

No one thought it was possible to develop smaller ones, and the market was consistently divided between screw compressors for air-cooled chiller applications and centrifugal compressors that used water cooling. Plus, the centrifugal devices operated at low speeds and low pressure ratios, limiting space to further energy efficiency.

All of these factors meant that the centrifugal technology was dominated by an extremely small number of large companies, which limited access for medium-sized and small regional entrepreneurs. This concentration fostered a conservative and captive industry with little motivation for innovation, with the result being that few new disruptive technologies emerged since Willis Carrier's first centrifugal compressor in 1922.

But Ron Conry's company, Turbocor, had something else in mind: the first ever oil-free compressor system. Magnetic bearing technology led to far less friction, removing the need for oil inside the machine and preserving the critical components, which made it exceptionally energy efficient. Integrated variable speed harvested the energy efficiency in part load operation, resulting in the first truly intelligent and connected compressor. Noise and vibration would be significantly reduced.

No oil would be used. Less power would be required. Fewer greenhouse gasses would be released into the atmosphere. Everything about the compressor was designed to be both environmentally friendly and less resource consuming. It was about saving the planet, one compressor at a time.

CEO Jørgen Clausen, heir to his father's company that began in a Danish countryside barn, would be the one to fight to acquire Turbocor at a time when there was serious debate inside Danfoss about whether the company should even remain in the compressor business.

The stage was set for two companies to collide, merge, and be reborn as the leader in oil-free cooling technology. One was a venerable veteran of Europe's cooling technology market that was angling to expand worldwide, which we can liken to an orchestra. The other, a fragile startup with a tremendous vision powering its small band of visionaries, closer to a jazz band.

Danfoss Turbocor would become the next step in sustainability, bolstering what's known as the "invisible infrastructure," the refrigeration and air conditioning that keep food safe, affordable, and available worldwide; that provide comfort to people at home, hospitals, and airports in any climate diversity; that enhance productivity in offices and manufacturing spaces; and that, in today's fast-paced computer science fields, cool high-performance chips in data centers, enabling AI transformation.

Perhaps fittingly, this revolution would eventually find its home in Florida, two hundred years and about 100 kilometers away from where Dr. Gorrie first hung blocks of ice from the ceiling of a medical ward in a futile attempt to battle disease. His persistence carries through the centuries, embodied in inventors like Willis Carrier, Mads Clausen, and Ron Conry, and is realized in the passion of owners, engineers, technicians, and customers of Danfoss Turbocor alike.

Scan the QR code below to learn more and view pictures.

# PART ONE—1994 TO 2004: TURBOCOR BEFORE DANFOSS

CHAPTER ONE

# Turbocor Is Born

Ron Conry, a native son of Melbourne, Australia, started his career working alongside his father-in-law, Roger, as an HVAC service technician. An inventor at heart, he moved to a smaller country town, married, and raised a family. He and his wife, Denise, were a couple who embraced their faith and lived family-centric lives. Early on, in order to treat their son Darren's auditory deficiencies, they moved to California.

Several years later, upon returning to Melbourne, Conry found that his business contacts had dried up, so he opted to strike out on his own and create his own business servicing air conditioning units and rebuilding compressors, specializing in large centrifugal machines.

Conry ran into complications trying to get parts from other manufacturers. He eventually decided to make his own based on the leftover components from the Ford machine he'd labored to repair. "It was enough for me to come up with a design and make a patent so I could cast a new impeller,"

Conry said, referring to a rotor used to increase the pressure and flow of a fluid in a centrifugal compressor. "But I couldn't find anyone to cast it. This one old guy had a foundry. He didn't want to do it, though. He just wanted to retire. So I offered to buy the business, and in turn they showed us how to make an impeller."

His workshop occupied a corner of an industrial building in Melbourne, where the concrete floor bore decades of oil stains from the compressors he'd rebuilt—ironic, considering what he was trying to eliminate. Salvaged components from Ford and GM lined the metal shelves like trophies from battles won.

Conry was still facing the difficulty of cost. The bigger impellers needed to run those compressors could cost in the tens of thousands of dollars. Conry hunted for economic solutions and, in the process, stumbled upon an aerospace design company that had exactly what he was looking for: an impeller far smaller than the industry standard.

At $95, it cost less than he might spend on an expensive dinner, yet it represented everything he wanted to change about the industry. "I'd never seen impellers that small before," he marveled. "I was used to paying twenty, thirty, or forty thousand dollars for one."

That planted the seed in Conry's mind that he could design an impeller like that one, capable of running an air conditioning chiller with far less energy and space needed. A few years later, he put that seed to use inventing a modular chiller system.

"Back then, and even today, conventional chillers were the size of a truck, placed in the basement or roofs of buildings," Conry explained. "Complicated, expensive. You could always cut up the unit and remove it in pieces, if you needed to, but to put it in was an ordeal."

Conry's pragmatism and innovation drove him to fix the problem at hand. His solution was a chiller that could fit into an elevator and then be stacked side by side to create a notably bigger combined unit as needed.

In 1989, Conry co-founded Multistack with business partner Roger Richmond-Smith, and they listed it in 1992. Conry produced thousands of modular chillers over the next several years, using modules that could fit in the elevator and be put side by side to replace the large chillers barred in basements of commercial buildings. He had to buy thousands of screw or piston compressors a year from the traditional air conditioner manufacturers while basically competing with them. Every purchase felt like funding the enemy. It was time to build his own weapon.

Moreover, repairing centrifugal compressors remained a frustrating process because most compressor failures were due to oil issues. This provided Conry much-needed work but sparked the idea that if he wanted to make his own compressor, he needed to get rid of the oil, make it smaller, and use a greener refrigerant.

At first, he chose R134a, which was becoming a popular new ozone-friendly refrigerant the automotive industry was adopting. However, he wanted to do additional research to find the best possible solution. He was spurred on, in part, by the scientific research into the depletion of Earth's ozone layer, which had resulted in the Montreal Protocol on Substances that Deplete the Ozone Layer, negotiated in 1987. This international environmental agreement was the first ever attempt to regulate the production and usage of almost one hundred man-made chemicals that were shown to be harmful to the ozone layer, a danger that especially impacted Australia. Cancer rates had shot through the roof, bringing the problem closer to home.

No one in the industry seemed to care much, except for maybe the auto industry. R134a was designed to resolve the issue of ozone depletion, but it created a new set of challenges, including hygroscopic oil and tremendous challenges with its miscibility. The Multistack modular chillers tried incorporating the new oil with the traditional compressors, and this led to many compressor failures. Around the same time, Multistack lost a major project in Singapore, forcing it to consider other options.

Ron took a business trip in 1993 in which he looked at myriad technologies and sat down with dozens of people, trying to find solutions that matched what he called his "mental specifications" for a device that could bring about real change. A visit to the Alaska pipeline uncovered a new kind of centrifugal compressor that led him further toward a compact, oil-free alternative. "I wanted it to cost one dollar, be as small as a matchbox, be 100 percent efficient, and be environmentally friendly," Conry said. "I had all these impossible goals. We didn't achieve them but came close, and at least we greatly improved over what was in the market."

One thing he knew for sure was that he wanted to round up people who had no experience in the cooling industry. "It's very hard to get tradition out of people who'd been involved for so long," Conry explained. "This was going to be revolutionary. The only way to get to it was to think revolutionarily."

He understood that eliminating oil from compressors was only half the battle. He knew he had to eliminate the friction of "that's how it's always been done."

Conry already had a certain person in mind: Yu Lin, an intelligent engineer and multilingual interpreter with a gift for electronics whom he had met in the 1980s when Conry was launching his Multistack modular chiller in China. They became fast friends, and Yu Lin moved to Australia, eventually earning a degree in refrigeration. Conry pitched his idea for what would later become Turbocor's iconic compressor.

In preparation for developing this compressor, Conry, along with an experienced lawyer, Doug Carter, filed the first patent for his concept on June 15, 1993. It was the first one to address the application of oil-free technology in refrigeration and air conditioning. Impressively, the patent was broad enough to be unclear as to what specific kind of oil-free technology was used—whether that be foil gas, hydrostatic, ceramic, or magnetic bearings—but narrow enough to establish a foothold in the HVAC industry.

This patent would later prove crucial. It established Turbocor as the undisputed pioneer of oil-free compressor technology for HVAC applications. No matter what claims would emerge from competitors in China, Japan, the United

States, or elsewhere, the record was clear: Turbocor was first. Conry is modest about what the patent achieved. "To me, a patent is just a road map of how long to do things," he said. "There's always other ways to do it. The real value is in giving investors confidence."

Frank Petersen, who served for many years as a patent engineer and later director of intellectual property rights, understands the dual significance better than most. He discovered just how good the Turbocor compressor patent was by researching the academic citations associated with it and certain other patents Danfoss acquired.

"If you write a scientific article regarding a patent, others will cite it, and then you count how many have cited this article," Petersen explained. "If there are only two citations, then it is probably not a revolutionary technology, but if there are maybe seventy citations or so, then it is a clear indicator that this is interesting or even disruptive."

He mentioned an award given to one of the inventors at a company he works for because that inventor's patent had been cited forty times in the past thirty years—an impressive track record. It wasn't an unusual number, but it was a high one. Then he looked at Danfoss's six thousand patents and discovered that the original Turbocor one filed by Conry in Australia had been cited 363 times since the 1990s, an extraordinary showing.

This gave the fledgling company a tremendous boost, allowing them to pursue funding to develop their vision. Another key to securing the future of that vision was Dr. Lin Sun, whom Conry and Richmond-Smith hired to become a member of the small, newly established team in March 1996. A former student and professor from Tsinghua University in China, Lin Sun joined Turbocor after he came to Australia in order to teach and research at Melbourne University. His doctoral dissertation—active bearing control and its role in various projects, including experiments with hydrostatic bearings and ceramics—was a godsend to Turbocor.

He saw the opportunity to do meaningful work that came with a lot of challenges and was so excited that he accepted the job even though it meant taking a pay cut compared to his salary at Melbourne University.

Conry tasked Lin Sun with evaluating the many options for oil-free enabling technologies, including the bearings previously mentioned. He would become the key leader in developing the technology and engineering that was Turbocor's lifeblood. He would also lead the transition of the people and organizational culture into what they are today.

"I knew industrialization would be a problem," Lin Sun said. "With that in mind, during the design phase, I purposely focused on manufacturability. Our team was exceedingly good at taking existing technologies and combining them. I had done that kind of thing since studying for my master's in China. It became about combining these cutting-edge elements into something new."

They eventually selected magnetic bearings as the best choice. Instead of metal parts grinding against each other, powerful magnets held the spinning shaft in perfect suspension, floating in air like a levitating train—no contact, no friction, no oil needed, no vibration.

Magnetic bearings in the centrifugal compressor, combined with variable speed technology, allowed for this massive shift in efficiency. The centrifugal compressors adjusted flow based on the system pressure. When the demand dropped, reducing the speed lowered both flow and power consumption dramatically. Even a modest speed reduction would yield significant energy savings.

This was quite different from piston or screw compressors, which are known as positive displacement machines. Those move a set volume regardless of pressure, and while speed changes affect the flow linearly, power savings are limited. These machines are designed to operate a full load; running them at partial speed leads to tremendous inefficiencies and major lubrication issues, such as the wearing down of parts, leading to a greater amount of lost efficiency.

In the Turbocor compressor, there is no friction between the moving parts and the lubricating oil because the oil is removed, which means there is no degradation of those same parts, nor degradation of the compressor performance over time. It's like a very efficient car that will run for fifteen to twenty years on 50 miles per gallon.

Using a high-speed motor allows the compressor to be built smaller as well. The basic way you generate pressure is by pressing on something, according to Lin Sun, who is now the chief technology officer for Danfoss Turbocor. According to physics, spinning is more efficient at creating pressure than pressing. The caveat with spinning is size and speed, because if you use lower speed (the traditional technology) to create spin, to generate an equal amount of pressure, the compressor will be massive.

To make the compressor smaller while creating the same pressure, it must spin faster, from 30,000 or 50,000 RPM (revolutions per minute), which is about fourteen times the speed of a conventional motor. Traditional bearings, however, cannot keep up with those speeds.

Using a magnetic field in a compressor allows the centrifuge to spin faster, thanks to the lack of friction. The team at Turbocor had been piecing together something truly remarkable. None of them had experience with compressors specifically. In any other industry, this would have been a liability. In this one, it was their secret weapon.

Electronics were another complex component that needed to be ironed out. "When we started, the electronics we needed didn't exist," Conry said. "We had to work with other companies to develop a chip with a fast enough computing capability, but the chips turned out to only be good enough to operate the bearings or motor, for example. We needed three chips just to run the entire compressor."

Of course, this revolutionary dream and technical acumen would only take Conry and his compatriots so far. Richmond-Smith, who was a brilliant strategist and fundraiser, undertook legendary work to keep Turbocor from bankruptcy. He was president and CEO of Turbocor during its challenging early years. He understood that to move forward with the compressor project, Turbocor would need to leave Multistack and raise capital.

Richmond-Smith secured initial government grants in Australia to fund proof-of-concept work, but those amounts wouldn't cut it. He spent countless hours flying around the world, making 286 funding pitches globally without any progress, before talking with Allan Myers and John Higgins.

"Ron had fascinating ideas about compressors and proved magnetic bearings' viability using an old computer showing the shaft position zero deviation after several months," Higgins recalled. This sparked his interest, which made him seriously consider investing.

Meanwhile, Allan Myers was vacationing in a beachside town of South Australia with his family and friends. Richmond-Smith, with whom Myers had previous connections, joined them.

"I recall very clearly, we were standing outside the house on a balcony overlooking the ocean," Myers said. "Roger had a glass of wine in his hand, which was not uncommon, and he told me about this Turbocor."

Richmond-Smith, when he finished sharing, asked, "Do you happen to have a lazy five million to invest?"

"Five million what?" Myers replied.

"U. S. dollars."

Myers thought that through. "Higgins and I will have a look at what you're talking about and decide then."

This led to Myers making a trip to the Turbocor workshop in Melbourne, where he saw the first magnetic bearing compressor in action. "It had been running for nine or ten months," Myers said. "There was this small screen representing where the moving part was. I asked, 'What size does that little space on the screen represent?'"

Conry reportedly answered, "That's about the breadth of a human hair."

Myers was astonished that, by magnetism, one could hold a fast-moving part in place for months with little to no variation. "That was absolutely amazing."

By 1997, the company had its first prototype. As Lin Sun recalled, "It was made from recycled materials from Multistack in Australia and PCB boards built at cost at his wife's universities, as the company scrimped and saved." That prototype compressor would eventually sit in the main office of Danfoss Turbocor, on display for visitors to see and to remind us where we came from. But it was only the beginning.

It wasn't long before Myers and Higgins agreed to inject the much needed $5 million into Turbocor. This became the first vote of confidence among investors outside Turbocor, helping the company limp through the first potential bankruptcy that threatened to shut its doors so early in its life. It bought them time, but they still lacked peace.

With several dozen engineers and their families now dependent on Turbocor's survival, Richmond-Smith found himself doing mental arithmetic late into the night, calculating how many weeks of payroll they had left, which suppliers could wait another month, whether his latest pitch would land before they hit zero. The weight of being responsible for other people's mortgages and their children's futures was heavier than any compressor he'd ever rebuilt.

Lack of funds severely impacted efforts to build the compressors. "We needed to buy electronics parts; we needed to buy cables; we needed to buy mechanical components," Lin Sun said. "But we were always running out of money. So at the beginning, we set up an account with the supplier, because we could not pay them right away. We had a struggle even to pay salaries."

Eventually, the supplier had had enough of letting Turbocor skate by until the next infusion of cash. They froze Turbocor's credit. The only way to get around that was for employees to use their personal cards to pay off the cost of the parts.

Manageability was another hurdle Lin Sun had to overcome. "We had different ideas coming from people with different backgrounds," he said. "Many had never done mechanics like I had. Again, we focused on the manufacturability and cost. At least one thing we all knew was that without looking at the cost, without considering manufacturing ability, there's no product."

He had the team sit down around a sheet of paper listing the many options they could start with in terms of parts and technologies integrated into the Turbocor compressor. One by one, Lin Sun scratched them off. Some had to go because they couldn't be reliably manufactured, others because of patent issues that existed. The dream was taking shape, but there were still seven painful years before Turbocor was able to sell its first unit.

Conry remains proud of what they achieved. He admitted that his strength was in picking people and motivating them with his own enthusiasm for the dream. "I think that showed through and became infectious," he said. "For an engineer, it was a great job to have, as long as we were able to keep paying them."

His idea to recruit those from outside the HVAC-R industry paid off, because the people who believed in Turbocor arrived without the baggage of "that's not the way it's done." They were willing to dream and to experiment.

Conry noted that Australia, while generous in terms of government grants for research and development, wasn't turning out to be a good source of investments long-term. Raising $3 million was much easier than raising $30 million.

Richmond-Smith pressed a great deal, but in the end, it was the Quebec government in Canada that extended the next significant lifeline to Turbocor: $20 million. The catch was the entire company had to pick up and move to the other side of the globe, into a province far removed culturally and environmentally from what the homegrown Australian endeavor was used to. They signed the agreement, and the next phase in Turbocor's short but tumultuous history took flight, quite literally.

Twenty engineers and their families moved to Canada, trading in their fans for heaters and their sandals for snow boots they didn't yet know how to lace properly. At Montreal's Trudeau Airport, they stood in hastily purchased winter coats that still had the tags tucked inside, their children clutching stuffed animals and asking when they were going home. These were folks who'd never scraped ice off a windshield, whose spouses were frantically flipping through

French phrase books, all betting their futures on magnetic bearings' oil-free dream.

Once settled in Quebec, the company's momentum was boosted by new talents on board like Robert Parsons, Steve Johnson, Francis Champagne, and Eugene Smithart, the last affectionately known by colleagues and friends as "Smitty," who left the HVAC-R industry giant Trane to come on board with Turbocor to assist with marketing. It was a huge coup for Turbocor, as far as Conry was concerned.

The fact that "Mr. Trane," as Smithart was also known, joined a startup company that was definitely not on anyone's radar drew tremendous attention. One of the other key actions was a demonstration model of the signature Turbocor compressor.

"[Smitty] said, 'Can't we put this thing in a suitcase and take it to the customers?' And I said, 'Well, the suitcase is going to have to be pretty big,'" Conry remembered, chuckling. "I told him it was going to be expensive, traveling with something that big, and he said, 'Don't worry about the money. Just do it.'"

"It" turned out to be a mobile showroom. Conry went out and bought a GMC 3500 pickup truck and a forty-foot fifth wheel camper trailer. Turbocor's people cut the trailer in half, built a meter-long extension, and installed a 60-kilowatt generator. The showcase would arrive in scorching parking lots from Phoenix to Dallas, where conventional rooftop chillers roared like jet engines. Sam and Brenda would power up the generator, extend the slideouts, and watch as a couple dozen customer representatives filed in.

The star of the show was the compressor—two of them, actually. One was a static cutaway model of the real thing, and the second was an operational unit. Condensing units and power supplies were installed where luggage would normally be stored at the front of the trailer.

"The customers were astonished by the quietness and speed of the magnetic bearing device. We'd open up a cupboard and here's the cutaway model ticking away," Conry said with a smile. "That's when they'd place their order."

Conry's daughter and son-in-law moved to Canada and took the show on the road, marketing Turbocor's revolutionary product. That marketing, combined with Smithart's effusive support for the compressor technology, is what brought Turbocor its first big customer. By early 2001, the company was producing only a few TT300 compressors, which ran with a 211 to 316 kW capacity and were both air- and water-cooled. (This compressor is still being built by Danfoss Turbocor over two decades later.) The difficulty in attracting sales, despite this marketing roadshow, was that very few original equipment manufacturers (OEMs) or the companies that dominated the chiller market were interested, although the curiosity of some service companies and contractors was peaking.

"Turbocor's model—high up-front cost, low lifetime cost—clashed with traditional HVAC players, who profited from aftermarket services," Richmond-Smith explained. "Industry authorities and OEMs were skeptical or resistant, fearing asset write-downs and disruption to existing business models."

The traveling compressor showroom stopped at Cypress College, California, in 2001, where unfortunately the compressor suffered initial performance issues due to the humidity's impact on the electronics. However, at the next site on their trip, a commercial building, representatives answering to the U. S. government had shown up, looking for the kind of technology Turbocor was producing.

The United States Navy was enacting a program with the goal of reducing their energy consumption. While that might seem surprising on the surface, consider this: the U. S. Navy is one of the largest consumers of fuel and electricity in the nation. A 2010 Defense Department report stated that "every day, the Navy consumes approximately 80,000 barrels of oil afloat and 20,000 megawatt hours of electricity ashore."

The General Service and Administration (GSA), which handles the management of federal facilities, including naval bases, was, and still is, looking for potential emerging technologies that it could then test and validate. The GSA and the Navy encountered a contact who had been exposed to the mobile showroom and, through mutual connections, was able to secure government funding to test the Turbocor compressor in a University of Southern California laboratory.

The test operated as follows: a traditional, oil-using compressor ran for several months, operating a chiller, and at the end of that span, the staff replaced that unit with a chiller using a Turbocor TT300 compressor. The results showed what Turbocor had been clamoring for customers to understand and raised eyebrows in the GSA and the Navy—the TT300 was between 35 and 40 percent more efficient.

As impressive as Turbocor's milestones were over the previous decade, it was this test that truly validated the company's technology in outside eyes. The validation hadn't come from the HVAC-R establishment but from an unexpected source, an end user: the U. S. military, an organization with no stake in protecting the status quo of the cooling industry. Sometimes the clearest vision comes from those standing outside the system—or those paying the energy bills.

It made such an impression that Turbocor compressor products ran quietly behind the scenes in chillers sold by our customers on the roofs and in the basements of U. S. Navy bases as near as Jacksonville, Florida, where Danfoss Turbocor is based, to San Diego, California; Honolulu, Hawaii; and the island of Okinawa in Japan.

For a small company that had struggled to put its ideas into a single prototype, Turbocor had made its mark on the map. More impressively, barely above the horizon, an exposition of worldwide HVAC manufacturers was about to bring the Australian inventors and financiers into contact with a stately Danish company that was on the hunt for emerging technologies that were energy efficient and environmentally friendlier than anything on the market.

Danfoss, expanding under the able leadership of CEO Jørgen Clausen, was about to come calling. What neither company knew was that their greatest challenge wouldn't be technology. It would be learning how to work together, learning from each other, and taking the best of each other's ideas. The result would be a brilliant group of innovators and entrepreneurs who would create a world-class global leader in manufacturing refrigeration, power electronics, and mobile hydraulics.

Together, they would form what today is a thriving oil-free community focused on one goal: the preservation of this one world we call home.

Scan the QR code below to learn more and view pictures.

CHAPTER TWO

# The Danfoss Gamble: "Vision—Accepting the Risk" and Betting on a Magnetic Dream

While Turbocor was moving itself to the other side of the planet thanks to funding from the Canadian government, Danfoss was at full speed on globalization.

China loomed on the horizon of the HVAC market. The giant nation was emerging from a long history of being primarily an agricultural country and beginning a period of rapid industrialization. The results of that decades-long effort can easily be seen today in China's domination of most manufacturing markets: the car you drive back from work through traffic jams, the couch you rest on after you walk in the door, the TV you turn on to stream your favorite show.

In 1996, I was part of the Danfoss global management team (GMM), which met every two years at that time to look at the company's strategic direction and make decisions about the company's overall approach to products and the market. I was the youngest member of the GMM and amazed to work with experienced and welcoming members. That year, we held the meeting in Singapore, in part because of its proximity to China. This was where the conversation turned to what role Danfoss should play in that emerging market.

Competition was heating up. Many companies in Europe and the U. S. were looking at how to establish themselves in China. Because of the unique challenges involved, they opted to try joint ventures with local companies and individuals to come alongside their operations while overcoming the cultural and language barriers. Cutting through red tape was even more difficult without making those arrangements.

Our team, however, did not recommend that path. While we acknowledged that the local connections and the joint venture setup could make starting out easier, questions remained. How were we going to develop an organization that fully embodied our core values? How were we going to build relationships with the local customers so that we could align our strategies and deliver long-term vision?

Instead, we recommended that Danfoss establish a presence in China that was wholly owned and operated by us. This turned out to be prescient on our part, because it wasn't long before we heard of many other companies that regretted signing on to joint ventures. They experienced headaches regarding the loss of control and created competitors that had strong connections with the customers, the very things we had been worried about.

Jørgen Clausen had just ascended to the position of CEO and was determined to establish China as the second home market. "We had recently received our business license to do business in China," Clausen said. "It was a very humble beginning, because we decided to go by ourselves."

In the ten years that followed, Clausen's leadership would help Danfoss secure success. He credits this success to the purpose the company and its employees shared. "They felt empowered to do it, and we fulfilled that strategy," Clausen said. "Such visions are extremely important to have, and today, people can still remember this."

In our view, people greatly underestimated how fast our Chinese counterparts were adapting to the international market. The three major players in China at the time, which included the now massive company Haier, were adopting and improving new technologies far faster than those in the U. S. and Europe. I noted this as I handled global accounts for Trane, Carrier, and York, which were by far the market dominators.

Much like the nimble visionaries running Turbocor, Chinese HVAC businesses weren't afraid to try doing things differently. They leapt from one technology to another, without taking the usual slower, linear developmental steps that might be considered necessary, and quickly became powerful, dominating market forces in the residential air conditioning industry. These technological leaps are evident today.

The first factory Danfoss established in China produced valves that controlled heating radiator systems, an elementary product, but gradually we brought more and more manufacturing to China. This eventually led to a second facility established in Tianjin, an area outside of Beijing that was fast becoming a kind of campus for these businesses. Initially, the primary benefit was operational cost advantage and the ease of exporting goods made in China to the wider market.

However, these developments were also about building our own competencies and expanding our footprint in the country so we could serve the market inside the country as it developed its infrastructure in a race to catch up with the West. Being a Danish corporation was an advantage in the long run, given that, as a smaller country, we were able to navigate red tape easily by sealing agreements in a friendly manner.

Danish culture is quite open, thanks to the fact that Danes are used to dealing with other cultures, languages, and differences in point of view. At that time, before things became solidified in terms of contracts, so-called gentlemen's agreements were the way deals were arranged with Chinese partnerships. Personal and professional relationships were the standard. The understanding was that if the partnership was no good, then there was no reason to stay together. If it worked for both parties, then there was an expected precondition that both would be, in fact, winning.

While this slow but steady market shift was taking place, technological dynamics were shaking the industry during the last few years of the 1990s and into the early 2000s. The Montreal Protocol and subsequent initiatives urged everyone to undergo a transition of refrigerants. As mentioned before, the threat to the ozone layer led to many discussions that ended with the entire industry coming together to ban these chemicals.

Danfoss was taking the lead by adopting several environmentally friendly refrigerants, including using R134a in addition to propane for the compressor in refrigerators and freezers, becoming one of a few participants in those early stages of the global shift. At the time, Turbocor was likewise switching to the greenest refrigerants, such as R134a.

The technology, too, was undergoing major change, beyond Turbocor's experimental magnetic bearings for its compressors. Up until this point, motors used in HVAC compressors were set at a fixed speed. They rotated at a certain RPM, and that was it. Variable speed meant that the motors that ran the compressors could adjust on the fly to the cooling or heating load of a building. For example, when the outdoor weather conditions changed, the compressor would slow down or speed up, adjusting to the fluctuation of the cooling or heating demand.

Danfoss had already been working hard at establishing its expertise in power electronics and becoming one of the largest producers of the Variable Frequency Drive (VFD) in electrical motors to save energy. This made Danfoss a natural candidate to see how both could be integrated into compressor technology. There were some challenges, however, when one considers the internal divisions.

By the early 2000s, Danfoss had its team of engineers in place to work on variable-speed compressors and marketing people to bring scroll compressors to the customers utilizing the new functionality. There was considerable discussion and debate surrounding how to integrate these technologies into the market and set Danfoss up at the forefront, overlapping with the company's initial expansion into Asia. Long days were spent in the labs, at the factories, and in the boardrooms, hunched over performance reports, their coffee and tea growing cold in their cups.

Meanwhile, Turbocor was settling into its second life as a Quebec-based company. Roger Richmond-Smith recalled that it was a tremendous shock when twenty people and their families moved from Australia, which in December 1999 was experiencing hot summer weather, to Quebec in January 2000, when below-freezing temperatures could be the norm.

This group of "visionary economic refugees," as Richmond-Smith called them, comprised most, but not all, of Turbocor. "We brought most of the key players," he said. "Some just didn't want to make the move. Some had other offers. We recruited three people from the outset, then four more, in Quebec."

They settled into a cold winter in North America. The Turbocor team was glad for the hard work testing and assembling the compressors—at least they could work up a sweat to ward off the chill. The irony of manufacturing these devices that would make air conditioning and refrigeration systems highly efficient while watching ice gather on the windows wasn't lost on them.

Meanwhile, engineers huddled at their desks as they toiled on design issues. While the prototype compressor had been a success in terms of drawing in potential customers and investors, challenges remained. Many of those issues revolved around the wiring and other connections in the compressor, separate from the magnetic bearings. The wires fed into the compressor via cables. Sealing those well enough to prevent outside conditions from interfering was a headache. Humidity was the enemy, rearing its head as condensation seeped, drop by drop, into the compressor's control systems. Electronics shorted out. Compressors whined and stuttered to a stop. Without the electronics, Turbocor couldn't accurately get the magnetic fields to balance, and if they couldn't balance, the rotor couldn't spin.

Those same electronic boards suffered overheating. Porosity in the aluminum casing meant the compressor couldn't be precisely sealed, either.

Richmond-Smith had his hands full keeping the people involved as passionate about their hard work as they wanted customers to be about the product's promise. French-speaking Quebecois who worked with Turbocor frowned at the boisterous Australians, with their funny accents and their strange way of doing things. His solution? Making sure honesty was the priority for all.

Outside the company walls, word was spreading about what this ragtag band of Australians was doing in French Canada. People were beginning to pay attention. The name "Turbocor" was raising eyebrows and prompting phone calls, and the company started to deliver some prototypes.

McQuay, a large chiller conglomerate founded in Minnesota around the same time as Danfoss, was already advancing energy efficiency through part-load optimization using their tandem centrifugal lubricated chiller. The system featured a highly innovative design and patented oil separation technology, offering impressive efficiency. However, it still relied on a costly oil management system involving pumps, controls, and fixed-speed induction motors.

CEO Ron Hallon and CTO Tom Watson recognized an opportunity to drive energy efficiency further by adopting oil-free magnetic centrifugal technology. "Our business model wasn't built around service," Watson explained. "It was centered on part-load efficiency. Turbocor was a perfect fit for our dual-head chiller compressor platform."

According to Watson, Hallon provided the visionary leadership to guide this transition, while Watson brought the technical expertise to implement the new approach rapidly. They both emerged as early leading advocates for oil-free technology in North America. Their bold introduction of the WMC dual-head Turbocor compressors marked the first major OEM deployment in the early 2000s. McQuay became Turbocor's largest customer for many years and played a pivotal role in shaping the early oil-free compressor community.

In 2002, Brian Evans was brought into Turbocor as the president, replacing Richmond-Smith, during its early startup phase, as the company sought leadership for commercializing their innovative compressor technology. Evans's deep ties to Trane were instrumental in shaping Turbocor's early steps. He leveraged his network to bring in key figures like Jim Schultz, former Trane CEO, who joined Turbocor's board; Eugene Smithart, former Trane sales and marketing;

and Dave Eber, Trane's former vice president of engineering, who consulted on the technology.

These connections played a part in building credibility and technical validation for Turbocor during its formative years, but it was extremely hard to find OEMs in the U. S. beyond McQuay and Multistack to adopt the technology.

"You know, what's interesting is that while we initially thought North America would be our primary market, Europe actually became the first to embrace Turbocor. The U. S. majors—Train, Carrier, York—they all made their own compressors, so they weren't looking to source externally and change their business model," Evans said. "But in Europe, especially in places like Germany and Italy, manufacturers were more open to buying compressors. That openness gave us a real opportunity. Axima in Germany and Climaveneta and Geoclima in Italy were some of our first big customers, thanks to Roger Richmond-Smith, who moved over there and started selling.

"That shift to Europe was unexpected, but it turned out to be a key move in getting the technology validated and commercialized. However, back then, we knew the technology had to be proven in the field, so we made a bold decision—because our volume was very low and the cost was very high, we sold compressors at a low price. We priced them competitively, going so far as to match the sales price of fixed-speed screw compressors. It didn't make sense financially at first glance, but the goal was clear: get them out there, make sure they worked, and build a reputation. You can have the greatest product in the world, but if it's not in the market, it's just a great idea sitting on a shelf."

York Zha was another one of those key individuals. He was a brilliant engineer working for Copeland, one of the largest compressor companies, based in Ohio. He'd often been hired out to do laboratory testing on new HVAC components.

Zha got his hands on the Turbocor compressor in 2002, and after he ran it, he was in shock. He could only stare at the results on the glowing screen, much like the University of Southern California researchers had done about a year before. This device, humming along quietly in a barebones workspace, was using far less power than it should have and was spinning faster than anything he'd ever seen before. And it was so compact, with fully integrated electronics and sensors.

Zha returned to China, where, at first, he remained in charge of research and development for this large American company, yet Turbocor remained in the forefront of his mind. He couldn't help but visualize how their compressor had the potential to make air conditioning units more efficient and sustainable. Looking out of his office window at the smog-choked skies that China's industrial rebirth was fomenting, he knew that oil-free, environmentally friendly machines were going to make a better future for this generation and the next.

Zha quit his job. He had absolutely zero money besides whatever he'd happened to save up, but he was entranced by Turbocor's vision. First, he raised money from his family, speaking in quick, excited tones as he extolled the magnetic bearing compressor over plates of steaming food. Then he sat before government officials in the city of Shenzhen, China, their posture perfect and precise as they listened with solemn skepticism on their faces.

Before long, he had received enough funding to start his own company. He hired the best and brightest from Tsinghua University and partnered with professors there to collaborate on case studies demonstrating the benefits of Turbocor's technology. They would, in turn, prove Zha's passion.

It took years to come to fruition, over a decade of hard work and dedication, but his passion was evident in the stock exchange designation for his company. BiXin Energy Technology became known as BSE, Believe in Saving Energy.

Later, he and a few other disrupters would emerge at the vanguard of the oil-free community in China, successfully convincing the market there that cleaner, efficient cooling systems were vital and cementing a booming business for the oil-free community in China.

Meanwhile, Turbocor, having unknowingly created an acolyte for its vision, was gearing up for an outing that had greater potential to make market inroads than its traveling showroom: the AHR (Air Conditioning, Heating, and Refrigerating) Expo in Chicago in January 2003.

I was the vice president of the global key account management for Danfoss, still handling three big North American air conditioning companies, the three biggest in the world. I visited most of the Carrier, Trane, and York facilities and headquarters all around the world, building the needed relationship with leaders and decision-making people. I also spent quite a bit of time in La Crosse, Wisconsin, which was the birthplace and former headquarters of Trane, to discuss how Danfoss could better supply valves, controls, and compressors that could be integrated with their products. That technology included an advanced version of scroll compressors running with new refrigerants, R410a, as the older R-22 was banned under the Montreal Protocol.

Trane, which had been acquired by American Standard in the 1980s, was undergoing large cuts to its research and development investments. Their R & D people were concerned by this diminished funding and the impact it could have on creating new technologies. David H. Eber was vice president of research development. Eber was an engineer with seventeen patents to his name and the key innovator who set up a "Skunk Works" of sorts within Trane. He thought Danfoss might have the courage to take greater risks. He pointed me in the direction of the Turbocor compressor at the AHR show.

We took our executives around to see the exhibit, which was already generating quite the buzz. Thousands of attendees milled about under the bright lights hanging high overhead. Booths were packed in next to each other, curtain to curtain, featuring innovative ideas from a planet's worth of HVAC industry representatives. The rumble of conversation filled the expo hall. There was a tangible energy in the air, like the static before a thunderstorm. AHR was alive with possibility.

When we reached the Turbocor booth, I was shocked by how small the compressor was. I was used to dealing with screw compressors, which is a different kind of technology for the same job. But Turbocor had produced a compressor that weighed 110 kilograms, far less than the 350-kilogram screw compressors I was used to. According to the data, it was 30 to 40 percent more efficient than a screw compressor. Another perk was that it took up less space than its contemporaries.

This was considerably more complex than what I was used to seeing. I thought it was brilliant how popular and accessible it could be made. Possibilities began to churn in my mind. My previous experience in Latin America, being in charge of all Danfoss products, had trained me in VFD applications and how they could save energy in turbomachineries and pumps. Images of how we could bring this technology into practical application arrived faster than what I knew to do with them. *We are looking at the future.*

Eugene Smithart was there, being absolutely fantastic at explaining the benefits of the technology. He was excellent at nailing down the key concepts for everyone. I remember he was standing there at the exhibit, speaking with our executives, while everybody was looking at the compressor. His excitement was contagious.

"Until that time, no one had really seen the concept of using oil-free compressors," Smithart would later say in a promotional video recorded for Danfoss. "The market reception for Turbocor was absolutely unbelievable. We were in the booth from opening until closing time at 6 p.m., and there were still people waiting in line to see this new compressor."

At the same time, the global air conditioning team was prodding Danfoss for the variable-speed scroll compressors, rather aggressively because they saw the opportunity. There were many, however, who told us that mixing electronics with a mechanical compressor would not be good in terms of cost, quality, and reliability.

So, when I saw the Turbocor compressor, with its power electronic systems fully integrated in place, balancing the magnetic field required for its frictionless operation, I knew it was what I had been dreaming of.

The name "Turbocor" remained on everybody's lips throughout the expo. Organizers had taken note beforehand, too. At an evening banquet, between the clatter of silverware and the laughter of people telling old stories and new jokes, they made a momentous announcement: Turbocor had won the Innovator Award.

After the show, I met with my team, and we reviewed what we had learned. They were extremely excited about the product, because that was the kind of technology we wanted to use to make inroads into the HVAC industry while supporting Danfoss's vision of creating environmentally friendly solutions.

Of course, there was continued hesitancy about investigating what Turbocor offered, because Danfoss was leaning away from building smaller compressors for home refrigerators and freezers. By the 1960s, over half of our business was dedicated to producing these units, and in the following decades, the market expanded to include many competitors. When companies shifted their manufacturing to China and Chinese companies began making their own versions in large volumes, the price began to drop. It became a difficult business to continue in, because the profit margins had sunk so low.

This became an emotional issue for Danfoss, to consider getting out of that market of "white goods" compressors, as they're referred to, and shift emphasis onto the larger, commercial models. We would eventually do just that, but around this time, when Danfoss was discovering Turbocor, the internal perception was that the compressor business was not good, so why would we want to invest in a new form of the same product?

The risk, in Jørgen Clausen's mind, would be well worth it. "I had come from research, and so I wasn't scared of things that are new, but I understood that bigger companies make very marginal improvements over time, incremental gains, with limited improvements, and they don't take risks. If you are a small player, you can afford to take risks."

Meanwhile, Turbocor was struggling to get the right combination of power electronics from suppliers. The industry at the time could handle rotors that spun inside the compressor at 3,600 RPM. But 45,000 or 50,000, more than ten times that speed? Managing that velocity required an especially unique type of power electronics. They had to be tailor-made for that kind of application.

Most other companies took one look at the daring specifications that Turbocor required and shook their heads. Too different. Too complicated. The volume was too small. Who knew if this fledgling company that had landed in Canada would survive much longer? One company, Semikron, was willing to take the risk, just like Clausen was willing to put Danfoss on the line to support Turbocor's vision. They created an entirely new package of electronics custom-made to provide the solution Turbocor needed.

Lin Sun took countless trips back and forth to Europe, nodding off during red-eye flights, navigating German transportation, to meet with Semikron engineers at their lab in Nuremberg. The collaboration was a win, resulting in an advanced, state-of-the-art electronics component that was certified in February 2004. Semikron became a true partner for Turbocor. Later on, in August 2022, Danfoss acquired Semikron, adding the last key technology to the vertical integrated mechatronic design and putting Turbocor in a much stronger position from a cost and technology point of view.

The vision of an oil-free community within the HVAC cooling industry was catching on among individuals like York Zha, visionary people in government entities like the U. S. Navy, suppliers and engineers and technicians that saw the potential of an unconventional compressor that could reshape the whole industry.

Despite its funding struggles, Turbocor was racking up wins—the kind of wins that, in the long run, would work wonders for its reputation. Then, in the spring of 2004, they added perhaps the most notable accomplishment: the U. S. Environmental Protection Agency (EPA) gave the company its Climate Protection Award.

Clausen challenged the Danfoss leadership to find a way to acquire at least 50 percent of Turbocor, but the initial discussions revealed that was too expensive given the company was still struggling. The back-and-forth continued for months, with face-to-face meetings in Denmark among the Danfoss executives. Scratchy voices echoed through conference call phones as they hammered out specifics of a possible deal with their Turbocor counterparts.

Myers and his family went so far as to visit the Clausen home in Denmark, where they were introduced to Bitten Clausen, the family matriarch and widow of founder Mads Clausen, with whom Myers's wife got along very well. This personal connection went a long way toward solidifying negotiations, though Turbocor had been seriously considering an offer earlier from a Brazilian company, EMBRACO. When the Brazilian currency suffered a collapse a few years earlier, however, that potential deal fell through, leaving the playing field wide open for Danfoss.

Per Have, the chief financial officer of Danfoss at that time, saw a benefit in forming a joint venture in which Danfoss and Turbocor, under its Australian shareholders, would each own half the company. He knew of similar situations in which companies, combining their stock through shareholder agreements, entered into a partnership managed by a board filled with representatives from both parties.

"When you come in, not as a majority shareholder, and stay focused on the business's profitability, it's easier to agree on a price," Have said. "You minimize risks to something that you already know is [heading] in the right direction."

Such shareholder structure brought the price down to a number that would allow Danfoss to build Turbocor with the cooperation of its Australian owners while leaving the option for them to withdraw. Eventually, Clausen suggested Danfoss buy into a joint venture, but only for the air conditioning portion of the business.

Doing so allowed Danfoss to kill the proverbial two birds with one stone: start the transition out of the market for white goods, which was proving unsalvageable, and into the manufacturing of commercial compressors for the hulking chillers that were used in hotels and government buildings.

Change is never easy. It soon became apparent that the ideologies and goals of the Danish owners of Danfoss differed from those of the Australians in charge of Turbocor. Both companies were successful and had successful leaders, but their expectations were at odds from the first day.

Scan the QR code below to learn more and view pictures.

# Part Two—2004 to 2008: Joint Venture and Turmoil

CHAPTER THREE

# Culture Clash and Cooling Dreams: The Joint Venture Begins

The first thing needed to operate the joint venture was combined leadership. Danfoss Turbocor, as the company would now be called, put together a board that brought in top leaders from both organizations, each with his own skill-set.

Turbocor had Allan Myers and John Higgins in place to represent the values that had driven their venture to take leaps ahead, even when the future was unsure. Danfoss elected Vagn Helberg, the president of refrigeration and controls, and Sven Ruder, the president of drives, to stand in for their interests. Mark Hopper acted as the executive secretary for the Australians, and Per Have, Danfoss CFO, served as the executive secretary for Danfoss.

There was a lot of excitement surrounding the deal, followed by immediate tension between the Australian and Danish contingents. Both groups came from cultures that were, compared to those of larger markets like China and America, smaller and more nimble. They were used to dealing with multiple cultures, both within their borders and in international markets. They shared a common focus on their people and in making certain they were prioritized, and both sides included experienced and successful businesspeople.

At the time, Ruder was in charge of the Danfoss drives division, a business with deep expertise in VFD. The forty years of expertise in his organization was critical because of the sheer number and complexity of electronics and power electronics involved in the Turbocor design. Discussions began immediately around each aspect of the joint venture company, such as how much would be invested in refrigeration applications, what size of chiller the compressors should be built for, and how to manage customers. There were a lot of opinions, as there should be. The difficulty was in reconciling those opinions around a clear, unified strategy.

"The psychology of the shareholders was key," Ruder said. "We can all look back now and smile, because things worked out very well, but at times the views on what needed to be done risked splitting the company apart. These conflicts ended up being good conflicts, but they didn't seem so at the time. They were very difficult."

Of course, how to handle the revolutionary patent Ron Conry had originally filed in Australia a decade before was another sticking point. When the joint venture formed initially, it was for the air conditioning side of the business. The application of Turbocor technology remained divided, as the air conditioning aspects continued as a separate business for a short time. This created a chaotic situation, with engineers having to scramble to determine which technology fell under Danfoss Turbocor's ownership and which did not. Referencing the original patent became a complex matter that resulted in misunderstanding.

These matters were further complicated by the fact that, twenty years ago, the board meetings were held in person. The shareholders met quarterly—sometimes in Europe, other times in the U. S., since technology for virtual meetings did not exist yet. Long hours spent in airplanes took their toll on the participants even before they arrived.

On the production floor in Montreal, well removed from the boardroom, Lin Sun was grappling with reliability issues for the compressor, making sure it could handle the humidity affecting the electronic systems. He was encouraging other engineers to be excellent at building the compressors and making sure they were technically sound while advocating for the oil-free revolution.

"We want them to get excited and try to present it to the potential customer [in a way that's] relevant," he said. "What's the meaning for them? We knew deep in our minds we needed to do this. It is a matter of talking with your heart, not just using the brain of an engineer."

With Danfoss joining the party, so to speak, this became an even easier concept to communicate with employees. "It has to be [about] thinking from [the] customer side, 'What does this technology mean for me?'" Lin Sun said.

But the most significant difference had to do with expectations, specifically the expectations of when Turbocor would first see profits. Myers and Higgins pressed for a three- to five-year increase from $40 million in revenues to $100 million—a gain of 150 percent. Higgins offered to fly the Danfoss Turbocor leadership team and their spouses to Australia for a dinner to celebrate the $100 million milestone. That part wasn't a joke, because they truly expected to see this exponential growth in such a short time.

The goal floored the Danes on the board. One must understand that Turbocor was deeply in the red at the time, losing around $8 million each year. The Danfoss executives, who had deep roots in their own company that had been in the HVAC industry since the 1930s, understood that change would take time.

Myers and Higgins shared an entrepreneur's view, one that emphasized quickness in both product development and profit. "They were not after long-term returns, but short-term," Conry said. "So there was always going to be conflict when you've got those views clashing on the board, but, in a way, that's what the board's about: taking different ideas and considering different structures."

Conry is convinced that if the board had been made up of Danfoss members alone, the situation would have probably been easier to manage, but the results would not have been anything like what Danfoss Turbocor became.

Jørgen Clausen's tremendous respect for Conry and his creation helped keep things together, a respect that hinged in part on Conry's visit to Denmark soon after the acquisition, which allowed the two men—the founder and the son of the founder—to see eye to eye and recognize their shared passion for oil-free cooling compressors.

Clausen had been keeping an eye on the board interactions and the creation of the venture from the CEO's seat in Denmark and later as the Chairman of the Danfoss board, but the joint company on the other side of the Atlantic Ocean needed someone to oversee day-to-day duties. The board decided that the GM of the Danfoss scroll business in the U. S. would be the man to manage the joint company.

He had spent five years as general manager of the Danfoss Commercial Compressor facility in Lawrenceville, Georgia, running the scroll manufacturing operation of Manuerope Compressors, bought by Danfoss in France back in 1993. That facility was the first Danfoss compressor factory in the U. S. He was a keen engineer with tremendous technical skill; however, his soft skills were not up to par for the complexity of the situation. A series of decisions and how they were articulated to the team and the board ended up shortening his brief tenure leading Danfoss Turbocor.

One of the many challenges this board had to grapple with was the lack of an appropriate manufacturing facility. The Turbocor building was a workshop, really, with offices for management and floor space where engineers could assemble the prototype compressor, toil on the complex electronics needed to run it, and precisely machine a shaft containing 142 parts, with a few microns' tolerance.

Where else could they go, though? Montreal, as elegant and cutting-edge as the city was in 2004 and 2005, lacked the manufacturing space Turbocor needed. Additionally, there was a shift in the government funding that had brought them to Canada in the first place.

Originally, the investment by the government made Quebec significant shareholders in Turbocor. At one point, they held 49 percent of those shares. When a new political administration came into office, right before the joint venture took place, they didn't see the benefit of holding shares in Turbocor and sought to sell their investment.

Myers and Higgins, who had invested an additional $5 million in infusions in the past decade, bought the government's shares, which meant they became the new majority owners and Conry and Richmond-Smith were the minority shareholders.

Another element demanding that Turbocor move was language. According to the local regulations, if a company surpassed ninety-nine employees on its payroll, all instructions and documentation it produced had to be written in French as well as English, adding tremendous cost and complexity to an already complex product and manufacturing process.

These restrictions were coupled with a desire to deepen connections with potential American customers. The North Atlantic Free Trade Agreement in the 1990s was getting pushback in the form of renewed calls for products to be "Made in America." This combined with the other pieces of the puzzle to form a clear picture: Turbocor needed to relocate to the United States. The question, then, was where.

The first CEO and president, along with Myers and Higgins, looked everywhere from Richmond, Virginia, to

Albuquerque, New Mexico, with Texas and North Dakota in between. Florida was attractive for a few reasons. One of them was the proximity to the MagLab, a national laboratory and the largest magnetic laboratory in the world, which had moved south from MIT in the early 1990s. It proved later that this was one of the best decisions ever made by the leaders at that time, to contribute to this magnetic ecosystem. The City of Tallahassee was also offering substantial financial incentives to bolster its manufacturing base. At the time, and still today, the state capital of Florida was dominated by a public service economy and higher education institutions, leaving little else to shore up its local tax base.

"Tallahassee was small," Myers remembered. "And Atlanta appealed because it was growing. One morning, at 8 a.m., I came into the office, and the phone rang. It was the president of Danfoss Turbocor, who said 'I need to tell you about a phone conversation I just had.'"

Myers had chuckled as the president and CEO recounted how Governor Jeb Bush of Florida had rattled off a series of reasons why the company should relocate to his state and the importance of being close to the MagLab. Eventually, Myers ended by saying to him, "We've already decided, we're going to Tallahassee. If the governor is making a call to help us decide, we're certainly going."

The deal they struck with Danfoss Turbocor was ideal: Tallahassee would build a brand-new facility for the company, including offices and manufacturing space, based on Turbocor's specifications. In return, the company would pay a monthly leasing fee, with an option down the road to purchase the building outright as it created jobs and brought economic development. The full benefits of being close to MagLab, as well as FSU and FAMU laboratories, eventually became clear. The huge draw of potential talents would start to build one of the strongest magnetic industries ecosystems in the world. But that was not well articulated with the Danfoss Turbocor team, nor with the board.

The downside was that the brand-new building, which was built in 2006 and measured about 55,000 square feet, was, in essence, an empty box. Turbocor had no furniture to provide for the offices or workspaces and didn't have the money to buy the newest versions.

This was when the desire of the Tallahassee community to work with businesses to welcome new industry became apparent. As soon as officials heard of the problem, they put Danfoss Turbocor in touch with a local technical school, Tallahassee State College, that was in the midst of replacing its furniture, meaning they had secondhand pieces available.

As this was taking place, news of the move was dropped onto the employees in Montreal like a hammer, with no discussions or explanations. The decision to let nearly everyone go, including Ron Conry and Lin Sun, could have ended the fledgling joint venture right then and there. At that point, most of what Turbocor had accomplished wasn't documented; it was what we call a true tribal-knowledge startup stage. There were some stacks of schematics and a few computer records here and there, but the bulk of the technical expertise resided firmly in the heads of the engineers who broke their backs each day to make the compressors operational.

The only consolation that Turbocor employees could hold on to was that their contracts would last six months from the announcement date. This time would be used to make that transfer of knowledge to new American employees.

One of those people being brought on board was Kit Kennedy, a mechanical engineer with experience manufacturing rocket engine components for Pratt & Whitney. Kennedy had earned her undergraduate degree from Florida State University and a master's in math before embarking on her career. At the time of the Danfoss Turbocor merger, she had found herself in a job that wasn't leading anywhere. It was her mother who pointed out a new company coming to town that was looking for engineers who would work on centrifugal compressors.

Danfoss Turbocor hired Kennedy in short order. They sent her and Ken Koehler, director of quality, to Montreal to learn all they could about what made Turbocor tick, from the tools they used to the broad methods of manufacturing.

While Danfoss was bringing Turbocor into its global family, CEO Jørgen Clausen was continuing the effort to globalize his father's company. The goal of expanding into China was thriving, no matter how difficult the market for compressors remained, and Clausen wanted to see for himself what Danfoss was getting into, albeit from a unique perspective.

During one overseas trip, he saw firsthand that many people had underestimated how quickly China was modernizing. When he returned to Denmark, Clausen insisted they find ways to accelerate the expansion of Danfoss into China. To that end, Danfoss engaged the services of McKinsey, a global consulting firm that was fast making a name for itself working with companies that wanted to build business in environmentally responsible ways.

This increasing pressure toward investment in China would soon send me on a fact-finding trip of my own, except to Montreal. Arthur Xu, a global key account manager of our China division (and today the president of Danfoss in China), accompanied me so we could assess the state of Turbocor.

Our global account team was energetic and excited by the Turbocor technology that we saw at the Chicago AHR show in 2003. We were so excited that we agreed that one of the best and most experienced team members, Scott Morehouse, a true believer in the oil-free technology, would be the first Danfoss person to join the new joint venture. We arrived in Canada with a business plan in hand that explained how to introduce Turbocor to the world and specifically the Chinese market. That report resulted in many startled faces looking back at us. I think the numbers we were looking for shocked them as they realized, for the first time, how much Turbocor needed to grow.

Ron Conry was excited by the idea, his dreams already big enough to propel him toward this. But Arthur and I found our enthusiasm tempered when we toured the workshop. It was the perfect space for entrepreneurial experimentation, but it was nothing like a production line. It could not churn out the thousands or even hundreds of compressors needed to make a dent in the market and satisfy the potential demand we were hoping to create.

I was coming from working with all the production lines in Danfoss and visiting most of the global accounts factories, with the understanding that manufacturing is about processes executed by people who focus on detail and execution. Turbocor, by contrast, was filled with brilliant people who would discuss their fantastic concepts with great passion, but there was little discipline. They would move from project to project in their work, a natural chaos for a small company that developed under the guiding hand of brilliant entrepreneurs and engineers committed to the oil-free vision.

In essence, Turbocor needed to grow out of their early startup phase. They lacked a detailed bill of materials, which should detail all the components used in assembly. They didn't have suppliers that were qualified to produce the motor and the power electronics. On top of that, it was apparent with this visit that the greatest challenge would not be the technology or manufacturing but the reeducation of the whole industry. We were not going to sell compressors, but rather, we were going to sell energy efficiency through a completely different business model—the "push and pull" strategy.

I knew that waiting would mean we could take the time to bring a better product to the market. The desire to leap ahead could cost us in the long run. I was determined to continue stirring up interest in oil-free technology, but only if it was a stable, viable technology produced by a company with a steady hand. It was equally important to have a clear disruptive strategy for changing the status quo.

Surprisingly, one of the larger Chinese companies couldn't wait that long. They were so eager to get their hands on Turbocor's oil-free compressor that they engaged in a bit of expo subterfuge. I and several other team members attended the China Refrigeration show in April 2005, where this company was premiering one of its new air conditioning chillers. We were walking the show with some of our global account team members. People were talking all around me about its new oil-free futures and high energy efficiency performance. That was when we looked up and noticed the Turbocor compressor was applied on a water-cooled chiller.

I stared at it. Others from Danfoss looked at me, their expressions equally astonished. "Who sold these Turbocor to them?" one of my colleagues muttered.

"I don't know." Nobody knew. So we approached for a better look, examining the compressor up close. On a hunch, I leaned forward and rapped my knuckles against the casing.

It was made of wood!

This company was so eager to get their hands on a brand-new technology they believed in that they were willing to show their new chiller design with the Turbocor compressor during a time we weren't ready to supply them.

But none of that excitement would come to fruition if Turbocor dismantled itself. The looming abrupt move from Quebec to Florida, coupled with the impending threat of termination for nearly everyone working so hard to make the dream of oil-free cooling compressors come to life, was poised to destroy Conry's vision.

Before the joint venture could become a true partnership, a true unified effort, the damage that was set in motion—admittedly by someone with the company's best interest at heart—would need to be reversed before it destroyed Turbocor.

Scan the QR code below to learn more and view pictures.

CHAPTER FOUR

# Turbulence in Tallahasee and the Great Leadership Reset

Imagine one of Turbocor's employees arriving in Tallahassee for the first time. It is hot and humid. The air is thick, like he's breathing water. He has left the cooler climate of Montreal for this swampy state he knows nothing about.

His wife and children are with him, tired from the travel, from trudging through customs and immigration at the airport. Their belongings will follow them in a truck.

This is the second time in less than six years they have moved—first from Australia to Canada, where they had to adapt their lives to fit with a predominantly French culture. Now, again, they must change—learn a new school system, make new friends, and adjust to American jargon.

He is wary, because at first he was told he was fired. He spent months teaching the art of crafting compressors to a young person who had no idea what the company meant to him.

Changes had happened swiftly. Danfoss Turbocor removed the first DTC CEO and president from his position after being dissatisfied with how he handled the move to Florida. Ron Conry and Lin Sun stepped in, with support from new Danfoss executives, all of them determined to assist the employees with their move, including arranging for green cards.

Before long, that Turbocor employee is met by a smiling real estate agent in a crisp outfit, hand extended to shake his. They've been hired by the City of Tallahassee and have been working with the family to find them the perfect place to live.

But what really makes a difference is the sight of two Turbocor employees waiting in the parking lot to greet them. These two had spent long hours in the Montreal workshop, laboring to fix electronics glitches and to painstakingly assemble the chiller compressors Turbocor was becoming known for. The smiles and hugs are heartfelt.

This was the spirit of the small families that sustained the Turbocor team members through a rough transition, a spirit of resilience.

Amidst the uncertainty of the move, decisions made by the first CEO and president of Danfoss Turbocor that were not well communicated led to replacing him with Juergen Suess as a transition president. Suess's job was to regain the confidence of the employees and the shareholders on both sides of the table.

Suess was a perfect pick to bridge this gap. While the board searched for the right person to be a permanent CEO, Suess worked to calm some of the disarray. He hailed from one of Danfoss's research and development divisions. A brilliant engineer, he could talk to anyone about the technology involved with Turbocor's compressors. He recognized it was a game-changer and wanted to do whatever it took to support that development.

But he also understood the kind of people who were creating the revolution. He was steeped in the culture of innovation and entrepreneurship—not only the project, but the minds that embraced the dream and the hearts that pumped with passion. One could say he was running on the same frequency as his Turbocor counterparts, and he had a natural chemistry with the engineers and workers.

Meanwhile, Conry and the other Turbocor stalwarts were strongly opposing the mass firing, and pretty soon, Suess made the same assessment they had. A true believer of the oil-free compressor, Suess would leave Danfoss in 2011 and become the managing director of one of the pioneers and the largest Danfoss Turbocor customer in Europe, Axima-Cofelly, today ENGIE Refrigeration in Lindau, Germany.

Conry was frustrated by this potential gutting of the company he'd worked so hard to build. But when he and Lin Sun looked deeper at the situation, they could see that fewer and fewer people were willing to make the move to Florida. They understood Turbocor was in danger of collapsing. Word got back to them that many on the team weren't going to go to Tallahassee unless he and Lin Sun did.

It wasn't long before Danfoss found a new CEO and president: Henrik Jørgensen. With him came Thorbjørn ("Thor") Hansen, acting as his vice president of operations.

He was a key player in the mobile hydraulics division, which meant he had no experience in the refrigeration and air conditioning industry, nor with the cooling compressor side of things, and certainly no connection to the oil-free technology Turbocor represented.

This, as it turned out, was a benefit rather than a potential hurdle. Jørgensen took up the mantle of leadership with no predefined concepts. A polite, friendly man, Jørgensen was considered the consummate people person, which fit perfectly with his new, primary role: to restore the damaged image of Danfoss that mismanagement had tarnished. Jørgensen began convincing those who had been told they were fired to move to Florida instead, in order to preserve their vast reservoir of technical knowledge.

Skepticism from the Turbocor employees did not help matters. Why Tallahassee? Why an area that had a small manufacturing base? In Montreal, Jørgensen and Thor attended meeting after meeting to address their concerns, facing pushback but doing their best to talk through problems, along with potential solutions.

It didn't take long to identify supply chain and operations as major issues, so Jørgensen brought Danfoss people in from Denmark and the U. S. to address those issues in preparation for the move. After all, Turbocor needed to be prepared to adjust from its workshop mentality to a factory setting, while still keeping the same creative energy and innovation for which they were already known.

One of the changes in leadership that went a long way toward overcoming conflicts was the addition of Kim Fausing.

Fausing was hired from Hilti, one of the most dominant electrical and power tool manufacturers in the world. Once again, there was a great advantage in having a new person join Danfoss from outside the HVAC industry. He had grown up in the same regions of Denmark in which Danfoss had thrived, which aided him in understanding the culture underpinning Danfoss's business operations.

At the time, I was reporting to Vagn Helberg, the president of our refrigeration and air conditioning business, whom Fausing was replacing. I met Fausing during his first days on the job in 2007 and toured the various global accounts we held, including Trane, York, and Carrier, three behemoths in the global HVAC industry.

He was immediately able to connect with the customers, I think because, unlike many of our excellent Danfoss team, he was less inclined to talk about the highly technical aspects of our products. Additionally, his experiences working around the world, including a long stint living in Japan, prepared him to interact with many different kinds of people and understand their cultures.

Fausing could visit with a customer and look at the business with fresh eyes. He understood that for us to be successful as a powerhouse of refrigeration and air conditioning, we needed to deal with companies like Carrier, Trane, and York, in addition to the emerging giant Asian firms in the international HVAC industry. The wholesalers and small manufacturers in Europe, while vital, would not be enough to make Danfoss a true global leader—hence the importance of having a strong presence in the United States and China instead of thinking of our market as "Europe and the rest of the world," as was the prevailing attitude within some Danfoss leaders at that time.

He would sit down with the CEO of a big company and ask, "How are we behaving as a partner?" More importantly, he would press them to explain, "What is your overall customer strategy? What kind of markets do you want to enter? How can we support you as a strategic partner?" It was all part of his goal to better position Danfoss as a company to serve customers in their plans.

"What I saw immediately from Turbocor was the potential," Fausing said. "It was something that didn't fly yet overall, but it was also clear that if we could make it work, this could really set the stage for many years to come within compressor technology."

Fausing would bring this vision to bear as he joined the board of Danfoss Turbocor, taking Helberg's place after he retired, which seated him with Sven Ruder. This addressed one of the concerns of the Australian shareholders, who didn't feel Danfoss was taking the joint venture seriously if they didn't put the most senior people possible, with fresh eyes, on the board. Having two Danfoss division presidents, Fausing and Ruder, working side by side—one with long experience in Danfoss and another who brought a new perspective—demonstrated the seriousness with which the company took this merger.

In the meantime, physical problems arose with the Turbocor compressor. Each compressor was built with four large capacitors, about a foot long and three inches in diameter.

Their role was deceptively simple: to prevent damage to the spinning shaft inside the compressor in the event of a power failure. If the shaft is spinning at 45,000 RPM and the power is abruptly cut, the magnetic field disappears in an instant, and with those magnetic bearings gone, the shaft spins out of control, wrecking itself and the inside of the compressor. The capacitors are designed not to prevent this power loss but to manage it in such a way as to avoid the damage. When the power goes out, the capacitor, which has been storing electrical energy, seizes control of the power supply to the electronics, maintaining the delicate balance of the magnetic bearings.

When working as designed, instead of cutting off immediately, the bearings peter out, allowing the shaft to slow down and make a softer landing. In that sense, the capacitors act as a complex battery. The problem Turbocor was facing was that the capacitors accumulate a lot of energy, all of which must be released rapidly to safeguard the compressor shaft. If the electronics that control the capacitors were to malfunction, there was a real risk that the capacitor would inadvertently release that energy when it wasn't needed. The technical term is that the capacitor would burst, but for all intents and purposes, it was an explosion.

Turbocor swiftly enacted a recall, because they found at least one case in which a capacitor had exploded. On the one hand, the external cover held back some of that force; on the other, the case shattered and flung around sharp debris. Obviously, this was a severe safety issue, one that needed a solution and fast.

A small group of engineers took on the problem with their typical gusto. They knew that in addition to correcting the error in the electronics that led to the power buildup, there was a need to modify the capacitor cover so that any explosion would be completely contained. To figure out what the threshold of strength was, they had to duplicate the same release of energy.

No one is really certain how they were able to achieve this, but the engineers created a cover reinforced so that, in the unlikely chance of a capacitor failure and a sudden release of energy, that energy would be directed down. The cover would keep the debris contained, preventing dangerous shrapnel from spraying all over the place.

Danfoss Turbocor put together kits with modified covers and new circuitry, along with specific instructions, which were then sent to the customers. They could, in turn, make the substitutions in short order. The timing turned out to be perfect, because the bulk of the compressors were still awaiting startup or had yet to be installed. Turbocor technicians went out to assist.

Within a few months, the repairs had been made. Ironically, the small amount of sales the company had made and the concentration of most of them in the U. S. made the modifications easier. The customers, who were mostly original equipment manufacturers (OEMs), were understandably willing to do this quickly and quietly, in order not to alarm the people for whom they were building chillers that would use the faulty compressors. It helped, too, that they believed in the Danfoss vision of oil-free technology. There was a great deal at stake, and they were willing to work toward a quick solution.

Murray Weightman, who was manager of Direct Energy business in Canada at the time, played a key role in making sure those compressor repairs took place quickly. "If you're using the product to build a business, you don't blame the product. You fix it," he said. "We fixed it so they didn't lose faith in us."

Weightman had seen every aspect of the compressor business as he worked himself up from apprentice to mechanic to chiller mechanic before becoming a branch manager and then a centrifugal division manager. He found out early on the benefits of Turbocor's visionary product when incorporated into chillers.

"The first job we did involved an eighty-ton Carrier reciprocating compressor in Woolworths in downtown Windsor [Ontario, Canada]," Weightman recalled. "It was ADX single circuit DX retrofit. We popped that out. We put Turbocor's in, piped it up, and it worked beautifully for years and years. They loved it because there was no noise and it saved a lot of energy. I think it would still be operating today if Woolworths was in business."

It was during 2006 that another major shift in the HVAC-R industry shook things up. The Japanese company Daikin acquired McQuay and, in doing so, not only supported the adoption of Turbocor technology, but also invested in developing its own oil-free magnetic bearing compressors for major capacity applications. Shigeki Hagiwara, known as "a compressor guru" with prior experience in oil-free technologies in Japan, moved rapidly to the highest levels of the organization and became the CEO of Daikin Applied Systems (McQuay), continuing the legacy of oil-free innovation and advocacy. This strategic move reinforced Daikin's commitment to oil-free technology and expanded the market for high-efficiency, low-maintenance centrifugal chillers.

Hagiwara had joined Daikin in 1981 and had discovered right away the issues with using oil in compressor systems. "I proposed to the top management of mechanical engineering that we needed to research oil-free options, in perhaps 1983 or 1984," he said. "Magnetic bearings, gas bearings, ceramic bearings . . . That was my first job and my first introduction to oil-free technology."

While Daikin's primary impetus for acquiring McQuay was to expand the market in North America, Hagiwara was delighted to discover their collaboration with Turbocor and was reconnected with his desire to develop oil-free compressors.

As the end of 2006 approached, there was cause for real celebration at the newly formed fifty-fifty joint venture of Danfoss Turbocor. The company had produced fifty compressors in two batches, and as the holidays approached, they were getting ready to ship them out—the first successful compressors shipped from the new facility in Florida.

Workers mounted the compressors atop wooden pallets. They bolted them into place, then fitted cardboard boxes onto them. The smell of wood dust and the chemical aroma of thick, hefty packing tape pervaded the warehouse. Heavy plastic straps were sealed around them. There were grins all around as people lined up the boxes for shipping, boxes bearing both the ornate red script stating, "Danfoss," and the bold, blocky blue word "Turbocor."

The bright word FRAGILE blazed on the other end of the box, with smaller black letters behind it in English and French. When the crew lined up for a group photo, the team all smiles, someone affixed white sheets that read, MADE IN THE USA and DECEMBER 22, 2006.

It was a huge milestone that could scarcely be imagined when one looked back at the humble beginnings in a Melbourne workshop on the other side of the globe. It's impressive to consider this was an almost entirely manual process. Turbocor was dependent on the people on the factory floor, and that was and is the beauty.

Kit Kennedy recalls the passion pervading the fledgling team that combined Turbocor veterans with new local engineers. "Especially when we started, the first five years down here, we were all one hundred percent engaged," she said. "We were building compressors that were making the world better for our kids, and we were proud to be doing it."

That is an important reminder that the oil-free community gathering steam in the early 2000s was thanks not only to the energy prompting wealthy investors, but also to the fervency of regular, hourly workers piecing together complex devices bit by bit. They were not only building compressors. They were building something that was changing the status quo for an entire industry, and they were only getting started.

Scan the QR code below to learn more and view pictures.

CHAPTER FIVE

# From Chaos to Clarity: Finding Footing in Florida

Shortly after Kim Fausing joined Danfoss as refrigeration and air conditioning president, he called a meeting of global executives in Austria. It turned out to have a far greater impact than previous leadership meetings. He gave out copies of *Good to Great: Why Some Companies Make the Leap and Others Don't*, a seminal book on leadership written by Jim Collins.

Fausing's goal was to encourage us to develop greater leadership behaviors that would drive Danfoss into the future. Those behaviors included personal humility, professional will, channeling ambition into the company, and looking in the mirror, not out the window (for failures). With these in mind, in his book, Collins emphasizes that strategy comes after assembling the right team, getting the right people "on the bus" and the wrong ones off before deciding where to drive it.

This meeting was a pivotal moment that positioned us globally where we are today and provided guidelines to follow in the years to come. Those of us in attendance gleaned a great deal of insights that we were able to use sooner than we expected.

During the first few years of Danfoss Turbocor's new leadership, the company was desperate to find customers. The need to move product and create a critical mass in volume was placed above all others. As a result of this strategy, Turbocor found itself selling its compressors to almost anyone who showed interest in the technology, sometimes at a price that didn't reflect the value of the benefit that was provided to end users.

Over half of the sales volume was for medium regional size OEMs like McQuay, Axima, Geoclima, Multistack, Smardt and Climaveneta, but no major global HVAC companies in the Western hemisphere believed in such disruptive technology, or at the very least, they were not willing to change their business model. The other half were mainly small contractors, doing retrofits that varied from project to project, who only needed one or two compressors at a time.

This model was quick but, ultimately, shortsighted. There was little follow-through. The relationship was transaction-based: here is the compressor; the customer pays the invoice; Turbocor makes the sales. And that was when things went correctly. Stan Zalik, our CFO at that time, had a difficult time collecting payment, since most of these companies were small and consistently dealing with cash flow problems, and we had minimal leverage after the compressor was shipped.

Customers were then using a new form of cooling technology that they were unfamiliar with, with a great deal of mechatronics, far different from the traditional fixed-speed screw compressor they were used to. They had the documentation they needed to tell them how it operated, and they applied the product. However, even if they had the same passion as Turbocor's people, the application challenge was still way above the capabilities of most contractors. The situation then became a problem of too many customers who, while they were buying the compressors in small volume, were not able to master the application of the technology. They were trying out new solutions, perhaps, or playing with a novel technology. They were willing to change the industry like Turbocor, but they would not be scalable and not have the global impact that we were looking for.

This created tremendous reliability issues and a potential credibility problem for the emerging company and technology. At one point, half of the compressors were coming back to us, either because the customer was unable to get the machine to work properly or because they were encountering some kind of technical issues, especially the integration with the chiller controller. Some of that had to do with mistaken assumptions about the technology. One could not simply plug a Turbocor compressor into a chiller unit and expect it to run flawlessly without adapting to a completely new centrifugal chiller controls strategy.

Let's say you have a car that's thirty years old. You remove the old but functional engine and swap in a newer, sleeker engine made by Ferrari—an efficient piece of machinery. This handles one problem, but if the car's suspension is shot, the brakes are worn down, and the transmission is failing, not even a good engine will save the entire machine. The car is simply not going to perform, and worse, the engine might suffer damage in the process. Similarly, the compressor would be blamed for not producing the expected chilled water needed.

This is what Turbocor was encountering. Customers were retrofitting decades-old chillers with new Turbocor compressors but keeping the old heat exchangers and controllers. To add to that, most of the chillers were full of oil and contaminants that didn't mix well with our compressors. The company spent a tremendous amount of time and probably 80 percent of our limited application engineers and service teams serving these contractors who were, it turned out, not the right type of customers. It was becoming clear that the technology could not be sold to just anybody.

For many years, our Tallahassee Turbocor warehouse was partly occupied by returned compressors living out the rest of their brief lives. These battered, somewhat humorous trophies had been through a lot, oftentimes with oil dripping from the inner workings because they had been connected to chillers by people who didn't understand what the magnetic bearings could do.

Richmond-Smith saw this misapprehension about the Turbocor technology firsthand. He was presenting the company's product during the planning of a new hospital in Montreal. Arrayed before him were the members of a review body headed by a senior academic from a prestigious university.

"After we made our presentation, he had this air of total triumph about him," Richmond-Smith said. "He surveyed the room and said, 'That's all very interesting, but let me point out that there's a fundamental flaw in all this.'"

A breathless hush overtook the room, as Richmond-Smith recalled. "Then he said, 'You've left out the oil pump.'"

The good news was that Turbocor was able to manufacture between 150 and 200 compressors per month after those first fifty were shipped toward the end of 2006. Somehow, the small daily output and close camaraderie of the company persisted through that first rocky year in Florida.

Kit Kennedy, who was working in production at the time, found that all she had to do was go out onto the floor and ask a few people for assistance if she needed it. The flexibility and familiarity meant that all were willing to pitch in.

But while the company entered a new era of stable leadership and fought to scrounge as many customers as it could, the market itself was teetering on the edge of a cliff. Global recession hit hard in 2008, cutting sharply into Danfoss's new, fledgling business.

The company as a whole was navigating rough seas. Danfoss customers stopped placing regular orders for small scroll compressors. Sales dropped through the floor. The hydraulics division, for example, found itself $500 million in debt. Sven Ruder, recently appointed as the president of the hydraulic business, had to make painful cuts to keep things afloat. Questions arose as to which business might not survive.

Turbocor, now settled in Florida's state capital, went through its worst financial performance, losing around $8 million a year. Shareholders on the board questioned how and when Turbocor could show return on investment if the global economy was to continue skimming the bottom.

They recognized, too, that Turbocor couldn't continue doing business as it had before, and throughout Danfoss, there was an understanding that change had to continue. Kim Fausing became the chief operating officer that same year, capitalizing on his leadership initiative to move out of the Danfoss refrigeration and air conditioning division. His natural talent for empathy toward customers and interaction with people within Danfoss would, the company hoped, steer it through the recession.

This appointment hinted at a brighter future for Danfoss Turbocor. Fausing had already spent time on the joint venture board and was familiar not only with the company's struggles, but also its strengths. His understanding would prove crucial for the company to maintain focus and its independence, even as it meshed practices closely with what had proven successful for Danfoss.

Still, it was apparent that the leadership situation at Turbocor itself needed adjusting. Henrik Jørgensen was doing a great job of smoothing things out, but his assignment was coming to the end and his family was getting ready to move back to Denmark. The change added to the perception of uncertainty in the higher ranks throughout the joint venture, though Ron Conry still held considerable sway. This was understandable given his role as the founder and the innovator who was driving nearly all the product development. It was his vision keeping the company energized.

The company was in need of reorganization, not because of any flaws, but because of the need to define what kind of customers Danfoss Turbocor should focus on and what should be the winning strategy. In addition, they needed to leverage the culture differences of large enterprises like Danfoss and a startup like Turbocor. Think of Danfoss as an orchestra, a body with a well-defined structure that assigned musicians to specific parts they were expected to play and play with absolutely perfection. They had their music. They were intentional when it came to playing the right notes in perfect time. For them, the sheet music consisted of playbooks such as the Danfoss Business System (DBS), created to drive continuous improvement in the business at that time.

Turbocor, on the other hand, was like a jazz band, full of innovative people who played by ear, relying on each other to improvise with little instruction. They didn't have much sheet music at all; rather, they had a fluid structure and brilliant minds to take advantage of.

This sort of freewheeling mode of operation with disproportional focus on the technology did not result in a clear strategy or allow Turbocor to work strategically with customers so they could share the oil-free vision Turbocor would use to change the global HVAC-R industry. We had discussed this exact difference in Danfoss, including with Kim Fausing and Sven Ruder. They understood. "Let's say we are going to make the changes," they said, "but there is something very interesting for you."

Both Fausing and Ruder, for whom I'd reported before, called me later with an offer. "We need you to help us with Turbocor. The company has tremendous potential but needs some direction. We need to create a strong vision and a strategic execution plan to help align the board and turn Turbocor around from losing money and find its full potential. We need someone who understands the global air conditioning industry and has connections with potential customers to partner with."

What they were asking was that I become the newest president and CEO.

It was not a request that came out of the blue, as Danfoss was restructuring the global accounts and air conditioning and moving in the product lines structure. It didn't take much to convince me, because I was already aware of some issues and had been following Turbocor throughout the past few years. I knew that it would not be an easy task, but the potential was huge and could have an impact not only in Danfoss but for the entire HVAC industry.

There was, of course, the lure of being the head of a joint venture, an independent legal entity, and a business with full profit and loss responsibility. It was a new challenge in and of itself. A tremendous opportunity to learn from an extraordinarily passionate group of people, including the board, and make a difference shouldn't be dismissed offhand.

This also wasn't a decision to be made lightly. I would consult my wife and my family. I would be jumping from the big boat of Danfoss into the little kayak of Danfoss Turbocor, and there were a lot of holes in that kayak. None of the kayakers could really see what was around the next bend in the river or if there was a waterfall around the corner.

The move to Tallahassee, in order for me to assume charge of Turbocor, would return us to a climate closer to what we'd experienced in the south of Brazil, but there was the looming question of what would happen if things went wrong. My main fear was for my family. They were concerned that they might have to move again if Turbocor didn't take off. I had a feeling that it was a great opportunity to join and serve the best team in the industry and to deliver something that was going to be remarkable. However, I would not do it if my family was not on board.

I brought Juliana and the kids to see Tallahassee. To be honest, they were shocked by the size of the airport and concerned by how small the city was in comparison to the sprawling Baltimore–Washington, D.C., area they had been used to for the past nine years. But Juliana embraced our philosophy again, and we knew that moving always had some risks involved.

"It's a chance to meet new people," she said, "to develop new relationships and be part of a smaller community and make a difference."

I had already come to Tallahassee to meet with John Higgins and Allan Myers for a job interview in October 2008. This was with the entire board, so although I already knew Fausing and Ruder, the two Australians were unknown to me. I immediately liked both men, very sharp and approachable.

A few days later, I received a phone call from John Higgins. I had the job of reporting to him as the chairman of the joint venture. It was strongly suggested for me to attend the annual Turbocor holiday party, which would take place about a month after the official announcement of my hiring, and introduce myself to the company.

The night of the party was chilly, as it turned out. Juliana and I never expected it to be that cold in Florida, which reminded us a bit of winters in Porto Alegre, our hometown, but not as cold as Baltimore. The party took place at the Tallahassee Automobile Museum, where Turbocor had rented a modest banquet hall. The building was packed full with marvels of engineering, cars of every shape and color shining under the overhead lights in pristine condition. The smell of motor oil and machine lubricants was in sharp contrast to the nice outfits everyone wore.

Attending the party was eye opening for someone coming on board to understand the company culture. The executives weren't the only ones there. Everybody who worked for Turbocor was there, including the folks on the manufacturing floor who were building the compressors by hand. It was an especially diverse team, representing over twenty-five different nationalities, and you could feel the power of collective intelligence. They gathered around tables, laughing and sampling food as they drank. Christmas tunes played in the background. I started to understand the core of the oil-free community, the passionate people of Danfoss Turbocor and the spirit of a small family of resilient and determined people, and I was starting to feel how honored I would be to be included on this team.

I recall Henrik Jørgensen, my predecessor in the CEO post, introducing me to the employees and their spouses. His thick, charming Danish accent proved to be a minor impediment, but funny, as he tried to tell everyone about the change that was coming. He indicated this to me and said they should welcome "a top leader from Danfoss," except his accent rendered the word "top" as "tough."

That earned some alarmed looks from a few of the employees. *Was another Danfoss tough guy appointed by Danfoss? Oh, no. Will this be another situation like the move to Tallahassee?*

Fortunately, both Henrik and Thor were able to put that worry to rest by clarifying immediately what he had meant. My previous visit to Turbocor made a difference, and my face was familiar to some. I got into many interesting conversations with the employees there. It wasn't until later that I learned why this party had started for the entire team. Apparently, during the last one held in Canada before the move to Florida, the manager was unwilling to pay even the modest expense of such festivities.

This was understandable, given the dire straits of the company's financials at that time. Instead, everybody chipped in to fund the party, and from then on, it became a vital tradition. Today, we honor and continue that tradition with a holiday party for all four hundred of our employees and their significant others.

That night, Henrik, Ron Conry, Lin Sun, and I walked the floor, and I met everyone from manufacturing staff to compressor engineers. The energy was electric. There was unity in the belief that we were transforming the HVAC industry, and I understood that we all had a clear and profound purpose.

That said, I was not coming in with a predefined agenda. My intention was to serve the Turbocor vision, to serve the Turbocor team, and to deliver meaningful results. To do this, I needed to figure out what Turbocor was like, how it worked, and bring together the best of both cultures, the orchestra and the jazz band. The first one hundred days would be my assessment period.

"I'm here to listen and help you guys," I explained to them as they gathered around. "I'm here to be one of your team. We are going to look at what's going in the right direction and what isn't, and then we're going to build a clear vision and a new strategy together to make sure that this technology takes its right place to transform the HVAC industry."

This received many smiles and nods, but I knew there were still lingering concerns underneath the polite expressions. I understood these concerns, because in one of my first all-hands staff meetings, I announced that Juliana and I had bought a house in Tallahassee and were settling in nicely. Afterward, one of the Australian engineers took me aside. He seemed incredulous. "You bought a house?"

"Yes, we did," I said. I was puzzled. "Why are you so surprised?"

He made a bit of a face. "None of the last four people in charge of this business ever bought a house in the same place we have the company." But then he smiled. "You're one of us."

I smiled back. "Oh, I'm not planning to go anywhere. I'm not going to be here for one or two years. I'm going to be here for a long time as part of this team." Past CEOs might not have shown their commitment, but I was all-in.

That settled things. I was absolutely convinced we had a great team with a great purpose, and together we would dig into Turbocor and find out what was holding the company back from great things. Meanwhile, the country and the world was scrambling to climb out of the pit of the Great Recession, which eventually ended in mid-2009. A reset would be difficult and painful, but it would launch Danfoss Turbocor onto a path toward a secure future.

We would have to work together if we were going to survive.

Scan the QR code below to learn more and view pictures.

# Part Three—2009 to 2013: Turnaround and New Strategy

CHAPTER SIX

# No Margin for Error: The Year We Almost Lost It All

If the holiday party was a welcome breath of fresh air, the new year of 2009 began with a deep breath. The advice that Kim Fausing and John Higgins had given me was ringing in my head: "Take the time during your first one hundred days to get to know the business and understand the issues. That is the time needed to look at the business with fresh eyes and have a clear picture."

That relatively short time was important because I did not have a substantial history with the company, nor did I have opinions about the way things should be run. The goal was to collect data and connect with people so that I could assemble the clearest picture possible of the entire company.

The approach was divided into three parts: get to know the team and its culture, understand the customers' needs and strategies, and learn enough about the technology underpinning Turbocor's products to understand what the customers were willing to pay for.

The first management meeting took place in a conference room at the Danfoss Turbocor plant in Tallahassee. The room was meant to be spacious enough to hold a meeting for ten people, with chairs set around a long table. Every single manager arrived, however, and it soon became clear that the room was a bit too small. The chairs around the table filled up, and people began sitting around the perimeter of the room.

Altogether, there were sixteen people who filled the room, all of whom reported to the CEO and president, each one of them filled with tremendous enthusiasm evidenced in their freely expressed opinions. Sixteen may not sound like a huge number of people, but it's important to keep in mind that at this point, Danfoss Turbocor had only about ninety salaried people on its payroll—approximately eighty people at the main facility in Tallahassee, plus a handful working in the field elsewhere in the U. S., a few others in Europe, and one representative in the Asia-Pacific.

Why such a top-heavy organization? Having key people the team could look up to and who could make decisions was important, but one of the primary reasons for the size of the leadership team was due to the chaos surrounding the initial plans to move from Quebec to Florida. The first president and CEO had left a vacuum in leadership as he flew back and forth between Quebec and Atlanta, where he lived. He was not in charge for long. Neither was Jurgen Suess or Henrik Jørgensen. Given their short tenures, although they had different results, it was clear that permanent, local leaders were needed.

Of course, Ron Conry and Lin Sun filled that vacuum throughout the rapid change in company presidents and during the move to Florida. Ron and Lin were the anchors, constantly holding everybody together.

I was hoping to change that by investing myself and my family in the new role and showing the Turbocor employees that I believed in the team and the company's potential. I wanted to stay for the long haul, hence the early announcement that the Schneiders were buying a house in Tallahassee. I saw the surprise on their faces and heard the murmured comments as it began to dawn on them that, perhaps, this time things would be different.

But before anything could change, we needed to rethink the leadership structure. The team was full of passion, yes, but it was too big and too unruly. During our initial meetings, it became clear that the team was operating without clear direction or strategy. Everyone had different priorities and ideas, which were good in and of themselves but not when they conflicted and slowed the progress we needed to make.

Obviously, this dynamic was perceived by the board as well, which led to their involvement, creating friction between shareholders. This was perceived as a lack of strategy and clarity by the customers and potential customers.

*We need to get this team down to five people, maybe six max,* I thought. The hardest thing that followed was telling the other ten or eleven people they would no longer be members of the leadership team.

This did not mean they were not part of building the new strategy or that we would let these key people go from the company, nor would anyone have titles stripped or pay reduced. Again, the size of the leadership team was a result of the chaos surrounding the big move. In order to convince the people who had been fired by the first CEO to come back to the company after that brief but confusing interim, those in charge at the time offered significant packages, prestigious positions, and inclusion on the leadership team. This was all in an attempt to entice them to return and keep Turbocor alive. It was part of the renegotiation process, and it worked, but it needed adjustments.

Those conversations had to be private and careful. "We really appreciate you," I would tell each one. "We really value your contribution. But from now on, while consulting all, we need to make very difficult decisions for which a small group is required to drive the discussions and actions with the board and customers."

There was thoughtful discussion around these changes, and I believe we did our best to articulate the new direction. Fortunately, everyone understood, and we didn't lose any players or their enthusiasm. Once the shift was made, we held our first meeting of the new leadership team in February 2009 at a hotel in Panama City. This small group, which included Ron Conry, along with Lin Sun, Thor Hansen, Stan Zalik, Ken Koehler, Uwe Seemann, and myself, put each issue on the table, every key item we needed to arrange within

Danfoss Turbocor. The outcome of this meeting was summarized on our first strategic milestone 2009–2010 Compass, called "Focus and Execution."

When the meeting ended and the first Compass was crafted, we looked at each other around the table and committed to having one unified voice: "The Leadership Team Voice." Naturally, the leadership team's members stepped up and started to create a better balance and execute the new strategy: Lin Sun for engineering, Thor Hansen for operations, Stan Zalik for finance, and myself focused on coaching the sales and marketing team to execute the strategy.

We shortly afterward called all coworkers in what we termed our "All Hands Meeting" and presented the Compass, explaining it in detail and adding a few suggestions from the team.

In this organization, everybody should have a voice, but we also wanted to ensure that all the voices didn't drown out a clear strategy and goals. Everyone should be able to identify what our goals and priorities were. We would run the business with the advice and support of the board, but it was our responsibility to run things on a day-to-day basis and deliver our promises to the shareholders and customers. The 2009–2010 Compass was the magnet that brought all Danfoss Turbocor teams together, from the people on the floor to the board.

Included in our responsibility was figuring out which customers we would serve, which products and projects we would focus on, and how to become profitable and generate our own cash. This may seem oddly simple. After all,

Turbocor was known for its oil-free magnetic bearing compressor. How many products could there be?

It turned out that there were thirty-two separate projects in various stages of contemplation.

Thirty-two? I did not understand. There was no documentation, nothing on paper. *How are we going to get to know all this stuff?*

Stan Zalik, our CFO, took me aside and mentioned a fellow Brazilian working in the research and development lab who might be a good source. We walked from my office into the bustling hive that was the Turbocor lab, which few people had access to. The space was filled to the brim with parts and people, all tinkering on one device or another.

The man's name was Jose Alvares, a brilliant guy with a deep technical background, especially in refrigeration, about to finish his MBA. He explained to us that Conry would pepper the team with numerous creative ideas for new designs, often scribbled down on a piece of paper, and the team would take off with them. The lab was pursuing many, sometimes conflicting, priorities at once.

It became clearer, then, why it was that Turbocor had dozens of products under development rather than a focus on a few proven concepts. As we continued our conversation, it became clear that the way product development was run had been left open for supreme innovation but was carried out with little discipline.

The lab, in effect, was still the same workshop of creativity it had always been, with no constraints on imagination, but no constraints on money and time spent, either. Small compressors, big compressors, different refrigerants for chillers, applications for uses in various businesses, and the list went on. For example, traditionally, our compressor has two stages of compression. We discovered that one of the projects percolating in the lab was a three-stage compressor that would provide lower, deep-freezing temperatures. Technically, it was doable, but it was a challenging project with questionable business to support it.

There was no clear documentation to answer some vital questions: Who are the customers we are going to try to serve with this particular new product? What's the budget? How about a business plan? No one had clear answers.

The leadership team was not fully aware of what was going on, and we were surprised with the findings. How do you start to build a prototype and build a test model if you don't even know what the customers are looking for? There could be many intriguing ideas, but if we were not developing those ideas with specific markets in mind, that was all they would be—ideas. As the leadership team continued our discussions, we decided to focus less on innovation and more on execution. We were dreaming up so many new things but finishing next to none of them.

This was the "jazz band" mentality that flourished. We had to cut the number of projects down to less than a handful that were in the pipeline and destined for our primary customers, like the improved TT300 and TT400 compressor, as well as the TT350 and TT500 under development. The company was still clawing for every dollar it could in the midst of a global economic crisis, and we needed to put an end to wild experimentation.

The innovation process requires some room to innovate, that much is true, but not unconstrained. We began adopting the two following principles:

- Innovation creates value that consumers are willing to pay for.
- Innovation is not complete until it shows up positively in the financial results.

We needed to learn how to balance innovation with execution, the understanding of whom the new product would be sold to, and the true value proposition of the product.

Our next goal was to review our customers. As mentioned before, about half of them were small retrofitting jobs, because Turbocor was desperately trying to find income. This inflated the total number of retrofitting customers to eighty-five. What Turbocor also faced was skepticism from the OEMs like Carrier, Trane, and the rest. They were curious about oil-free compressor technology but unwilling to risk changing their business models. And the smaller retrofitters who were buying only a couple of compressors per year were having difficulty adapting to our revolutionary technology. They were excited about the technology; however, business with them was not scalable.

The leadership team concluded that we needed to focus on building the oil-free community, a group of believers like OEMs, end users, engineering firms, and universities, on top of early believers. The oil-free community would create a push and pull dynamic, driving a significant number through the OEMs, which would aid us in standardizing applications and optimizing product performance as they mastered the technology, improving reliability and chiller efficiency.

Around mid-2009, we convened a call with these retrofitting small contractors. To continue the automotive analogy I referred to previously, these were the people who had been putting brand-new engines in old cars. We told them, "Thank you very much for believing in the oil-free compressor technology. We love your business, and you love this technology. You're doing a really great job promoting it. But from now on, instead of buying an engine, you're going to buy the whole car."

In essence, we were not going to sell Turbocor compressors to those contractors in the service industry. They would need to replace entire chillers rather than just the single part—a part that, in many cases, was ill-suited to running a chiller that was already decades old.

The contractors were upset at the sudden change in how we were doing business. Ron Conry was disappointed, too, because he came from the service and retrofitting field during his early years in Australia. His passion for the people, while admirable, put him at odds with our direction, but we explained to him that if we didn't let those customers go and focus on a few OEMs that could scale up the application, the company would not survive.

In the end, he did understand, but he was right: some of them were not happy. A big advocate for Turbocor and energy efficiency, Charlie Fletcher, was one of those retrofitters.

His California company, EMCOR MESA, was one of the largest service contractors and installers on the West Coast. He was a true visionary when it came to energy efficiency and innovation, but following this discussion, he was so upset with us that we didn't speak for a few years.

Later, we happened to run into each other in the corridors of an HVAC expo. I don't recall who offered first, but we wound up sitting down for beers together. Charlie admitted that in the long run, his company had fared better by replacing the whole chiller instead of the compressor by itself. DTC's decision had steered his business and many others in a better direction.

Charlie had installed hundreds of Daikin, Smardt, and Multistack oil-free chillers on the West Coast and became a force to be reckoned with in the industry. He was a true believer and advocate for energy efficiency and innovation and a key leader in the oil-free community.

Redirecting from the service contractors to the OEMs was the first step in our strategy to create demand, or what we called "pull strategy," but we were still too far from the end users, the real beneficiaries of our value proposition, the ones who would pay for the next fifteen to twenty years of the electricity consumed by chillers and their maintenance.

Another issue was the desperate situation to increase sales volume of compressors in the few OEMs on board. With the lack of sales and marketing experience in the organization, Danfoss Turbocor was selling amazing technology at a ridiculously low price. Actually, they were selling to some OEM customers at the same price as the fixed-speed screw compressor. Back to the analogy of the car industry—due to the lack of commercial experience, they were selling a BMW at the same price as a manual geared Honda Civic. We needed to understand and demonstrate the value of our technology and a clear value proposition and reeducate the industry.

We took the template developed by my previous global account team to calculate the right value of the Danfoss Turbocor product, and to do it right, we needed a good engineer with an MBA. That engineer was Jose Alvares, the one I'd encountered in Turbocor's R & D lab. We promoted him to product manager and gave him this task.

It became clear that the greatest benefit was the energy efficiency gain. Joe Alvares formed a partnership with Scott Morehouse, who was familiar with the tools from the Danfoss global account, and assembled a small team. Alvarez and his team used various software tools to simulate thousands of commercial buildings in different climate conditions and occupancies, comparing the energy consumption between chillers with oil-free compressors and the traditional fixed- and variable-speed screw compressor.

It was tremendous work, resulting in the equivalent of a PhD thesis with a lot of hard data showing that we should be charging at least 25 percent more for our compressor than the traditional ones. The payback was only two to three years in chillers that were running above three thousand hours per year, assuming the cost of electricity was above fifteen cents per kilowatt-hour.

The team created a simple intuitive tool in the form of an app for mobile phones to play with the simulation, allowing thousands of people to better understand the value proposition of the oil-free systems. Together with our selected OEMs and with these tools, we began educating the industry about the total life cycle cost of chillers, including energy efficiency and maintenance cost.

We needed to deploy the new price strategy, and by the end of 2009, we did, with a big customer in Europe who was an Italian OEM, Climaveneta. Carlo Grossi was the CEO, leading a great team, including Mr. Sormani, the R & D leader, and Maurizio Marchesini, the COO. All three were believers in oil-free technology, great promoters of its benefits, and among the earliest adopters of the Turbocor compressors in Europe.

Our visit to his team later in 2009 did not go much better than when I first informed Charlie. We had gathered around a table for a great Italian dinner, complete with wine and plates piled high with enough food to make one drowsy. Climaveneta was a leading company, the largest chiller OEM in Europe, so we had to keep them buying the compressor, but we had to be clear with them.

"We have a problem in Turbocor," we told Carlo and his people during the dinner in Bassano del Grappa. "Namely, that our product is underpriced. Our unexperienced people were, in the past, desperate to have units running in the field, and it is not right to sell Turbocor at the same price as a fixed-speed screw compressor. Turbocor is 30 to 40 percent more efficient, with lower noise and weight. As a result, we are charging at least 25 percent lower than what we should be.

Starting in 2010, we need to increase by approximately that much."

It's safe to say the dinner exploded. Lots of Italians yelling simultaneously. Lots of gestures. Emotions ran high, with accusations of "You can't do this!" being tossed around. It took a while to lower the temperature and a few months more to negotiate before they accepted the new price.

Several years later, I had a follow-up encounter with Carlo. We visited over a glass of wine, and Carlo asked me if I remembered that fateful dinner. "I do remember," I said. "I know it was needed."

"Well," Carlo said, with a smile, "we were just waiting for you to come to us and tell us you were correcting your price. We all knew the old price was wrong. How could you be selling such a beautiful technology at the same price as a fixed screw? It was not right!"

I was impressed that he could be that open, but of course, that is the benefit of maintaining these relationships with customers who were big supporters of oil-free compressor technology. At the same time, it was clearly part of Turbocor's weakness, a focus on the beauty of the technology they had created to the extent that they didn't yet fully understand what customers were willing to pay for it—hence our redirection to fewer products with a clear value proposition and fewer customers.

The mission of the oil-free community was clear. We needed to educate the HVAC industry together—the OEMs, engineering consultants, university professors, industry associations, and end users—about the total life cycle cost of oil-free air-cooled and water-cooled chillers, and we knew that was not going to be easy.

Immediately, we jumped on a great Danfoss initiative in the U. S. called "Danfoss EnVisioneering," a strategic initiative that blended engineering innovation with visionary thinking to address the evolving challenges of energy efficiency, climate impact, and sustainability in the HVAC-R industry. The program started in 2006, and we saw it as a perfect forum to create the visibility needed for Turbocor.

Many showcases about energy efficiency were presented and debated during the symposium through the years. In 2010, it established the Envisioneering Award, with an independent panel to select the best case in energy efficiency and sustainability. Demonstration cases with real hard data with chillers with Turbocor were showing and confirming the value proposition, with savings from 36 to 61 percent:

2011—Multistack

2012—Smardt Chiller Group

2014—Daikin (McQuay)

2015—EMCOR Mesa

2016—Smardt Chiller Group

2018—Multistack

2019—Artic Chiller Group

2023—ThermalWorks

2024—Thermal Works

This was the push and pull strategy in action. We were getting visibility in North America, with a tremendous collaboration with Danfoss communication director Lisa Tryson leading the activity.

Now President and CEO of Smardt Chiller Group, Richmond-Smith said, "We pretty much had to change the whole business model used by the whole industry, especially regarding the total life cycle cost of a chiller. Traditionally, they were oversized at 20 percent, at least, above capacity."

This was done to meet peak demand with the conventional safety margins, which happens once a year for a few hours during the summer, but doing so meant these oversized units were operating inefficiently during the rest of the year and for their entire lifespan. "We kept pointing this out, but a lot of people refused to accept this fact," Richmond-Smith explained. "It took a long time to hammer away at this misconception with governments, customers, and even regulatory bodies like AHRI."

This further emphasized that the business model had to be sold before the technology could gain traction.

Kim Fausing, COO at the time, knew those relationships were vital. "We needed to find ways to go faster, because our big worry was the customers, not the company," he said. "If you have a new technology, you can't simply develop it yourself. It's not that you sit in a lab and you develop something, and then you have a product and go to the market. It's something you do in relation to the applications of your customers."

Fortunately, the sales team at Danfoss Turbocor was starting to build deep customer intimacy with a few key customers, who would later contribute to that success. "If they would not have thrown themselves at this, we would not have been where we are today," Fausing asserted.

It was obvious that we had the best people in the industry from a technology point of view, and they did not need my support, but I was eager to learn the basics. Daily one-hour meetings with Ron Conry and Lin Sun were my boot camp, during which we would carefully study the details of the compressor's inner workings, how the magnetic fields kept the shaft levitating, and how the compressor was controlling the speed and capable of delivering such high energy efficiency.

Lin Sun, a former professor at both Tsinghua University and the University of Melbourne, possessed a deep passion and an innate talent for teaching. He had a remarkable ability to simplify highly complex topics, making them accessible to all. He often said, "If you cannot explain an issue in a simple way, it means you do not truly understand it."

Despite our cordial working relationship, there was an undercurrent of unease. Conry was having a difficult time, not with any one person but with the direction in which the company was moving. The shift away from freewheeling tinkering and the elimination of small service and retrofit customers had hit him especially hard, because they embodied everything he felt made Turbocor special. This was true, but Turbocor was becoming Danfoss Turbocor, a blend of jazz band and orchestra in practice as well as on paper, and without necessary changes, it would not last in either form.

Ron's urge to keep innovating clashed with Turbocor's need to develop stable procedures. I proposed appointing him as a permanent consultant, advisor, and influencer who could travel the world promoting oil-free compressor technology, training people in how it worked and what a tremendous difference it could make, and thus create demand as part of our pull strategy.

However, Ron still strove to engage in his own projects, but to do so, he needed approval of the fifty-fifty joint venture. In the end, there was too much oversight and not enough freedom for him. "I couldn't cope with that sort of bureaucracy," he said. "I'm just a different kind of person, an entrepreneur. I've worked for myself since I was twenty-two."

I again cautioned Ron against striking out on his own, but wanted to make sure he could do so if that was his final decision. We spent six months discussing scenarios for how Ron could start his own company all over again and have the rights to utilize Turbocor technology from us. We were trying to balance his need to innovate and the dire financial straits Turbocor was still facing, in the midst of our major changes.

We emphasized two conditions: Danfoss Turbocor could not take any financial risks, and we could not have any distraction within the organization from Ron, because as the founder and role model, we knew people still working for us would be willing to drop what they were doing to assist him.

We treated the situation with extreme care, because everybody knew Ron in the industry. Those connections were real, and we wanted everyone else to understand this parting of the ways wasn't about animosity toward him.

Things could have gone terribly wrong. I spent many nights in bed, staring at the ceiling, tossing around ideas about how to do this the right way. *It has to be good for Ron. It has to be good for Danfoss Turbocor.*

When we put the framework together and we presented it to the board, they immediately gave their total support. They understood, as we did, that Ron is a wonderful guy, and a brilliant innovator and entrepreneur, who thrives in the absence of limits—yet limits were what was needed to save Turbocor.

This separation was the hardest part of that critical year, 2009, that came close to seeing Turbocor's potential derailed. But by urging the jazz band to take up some orchestral arrangements while still allowing them freedom to improvise, we stretched ourselves to remake the company into a disciplined yet innovative proponent of a visionary technology.

Scan the QR code below to learn more and view pictures.

CHAPTER SEVEN

# Execution Over Everything: Patent Wars and Profit Plans

It was reassuring to spend most of 2009 focused on building and communicating a new strategy, "Focus and Execution," with an emphasis on "push and pull," a few customers to serve, and a unified team to deliver our promises. However, difficulties had emerged right before I joined the ranks.

The original Turbocor patent filed by Ron Conry had served product development and innovation admirably for many years, but we heard some discussion that this same patent faced potential infringement from other HVAC companies. These companies had seen our potential success and were determined to develop oil-free compressors, and we needed to make sure that our patents were respected.

Several companies across the HVAC industry were exploring technologies similar to Turbocor's, including some with whom we had trusted and long-standing commercial relationships.

Within the first few months of my tenure, we set up meetings to discuss the matter directly as partners. We had established a good rapport, so I called him up and explained that Turbocor had reason to believe that certain new chiller platforms might overlap with our technology. There were many ways we could resolve the matter.

The discussions were constructive, and we continued the dialogue over the following weeks. One way these patent infringement matters are often resolved involves royalties. The company using the patent agrees to pay the patent holder a certain amount for using the designs in question. We followed up with meetings in Washington, D.C., to explore possible commercial arrangements that would allow the technology to be used under appropriate licensing terms.

The position presented claimed that there was no infringement.

No discussion. No consideration of royalties.

We were taken aback, but it did not take long for us to firmly state our position. We explained that Turbocor was a joint venture with its own shareholders and that we, therefore, had a responsibility to protect the company's intellectual property.

It was clear that the discussion had moved from a constructive business dialogue to a formal legal position. The shift was unmistakable. It was a far cry from the back-and-forth discussions I'd started with. But the main value of small start-up companies like Danfoss Turbocor is the patents. If we do not protect the patents this company created, the value will diminish.

Also on our minds was the protection of our selected customers who were making commitments and investments in the oil-free community. We were determined to protect our IPs and not let our customers down.

The meeting ended curtly, with neither side budging.

Once we returned to Tallahassee, we gathered the leadership team and evaluated the overall situation. We agreed on the next move and called for a special meeting of the DTC board. There was, unfortunately, only one course of action: to pursue legal protection of our intellectual property.

I thought there might be some debate, but all the board members gave their approval. There was no question about it.

I found out later that the matter had to be raised with the Danfoss board as well, and I was happy to learn that they saw this issue in the same way. In truth, I was not surprised. Protecting innovation through patents has long been part of the Danfoss DNA, going back to our founder and the company's earliest days, and defending intellectual property was widely understood inside the company as a matter of principle. If we did not protect the current Danfoss Turbocor patents, we were going to have a problem supporting and defending our other patents in the future.

The matter eventually entered a formal legal process, which is a matter of public record. As part of that process the parties went through mediation. We will not examine the specific outcomes in detail here, but suffice it to say that the result confirmed the strength of Turbocor's intellectual property. The entire ordeal took eighteen months from start to finish and was a great lesson on how the legal system in the U. S. is efficient and relatively fast but expensive.

The outcome strengthened Turbocor's patent position and sent a clear signal to the industry that we would actively defend our intellectual property.

Some observers felt that defending patents so firmly against customers and industry players could complicate relationships, but Danfoss leadership remained clear on the principle: Intellectual property must be respected if innovation is to thrive. Respecting IPs is a question on principles, and property rights must be defended consistently even if you are challenged by valued customers, because if you fail to defend them once, you weaken them everywhere.

In the meantime, we had a call from a representative for McQuay, a Turbocor customer that became part of the Japanese multinational Daikin, which became the world's largest air conditioner manufacturer. "We have seen some noise on the market about this lawsuit, so we need to sit down and talk it through," he said, and I was happy to do so.

I met with Ron Hallon, who was the CEO of McQuay at the time and one of the early believers in oil-free technology. He understood our position and referred me to Daikin Japan. We dealt directly with the CEO of Daikin Applied, Mr. Hagiwara.

When we heard the name, it was quite a relief, as I knew him from previous meetings regarding scroll patents shared by Trane, Daikin, and Danfoss. During our conversation with Hagiwara's team, we got up from the table and moved to the white board mounted on the far wall. Dry erase markers squeaked against its surface as we wrote out various scenarios, jotted down many ideas, and pondered the variables involved. McQuay (Daikin) was buying about 20 percent of our total volume in 2010, directly from Turbocor, as compared to the previous HVAC company, which was a Danfoss customer but not buying Turbocor.

By the time we finished, we had arranged what I considered an elegant solution. It was a supply and cross license technology agreement that stipulated Daikin was obligated to buy a certain number of compressors from us each year, a number that was a substantial increase from the year before. The prices we agreed on took royalties into account, which then meant Daikin was free to put our patent to use in their 400-plus-ton compressors. We shook hands and set about drafting the requisite documents that were fully approved by the Danfoss Turbocor board and were signed by Hagiwara, Hallon, and myself.

It was a significant relief compared to what we had gone through with our previous patent fight. The Japanese business culture had a deep respect for the integrity of patents, and they preferred to strike a balance that was beneficial to both parties. A couple of months later, another Japanese HVAC company contacted us, seeking rapprochement based on that same respect. In that instance, they told us flat out that they were stopping development of their own compressors, so there was no remaining question of patent infringement.

Our reputation had been solidified. Other companies understood the lengths to which we would go to protect our intellectual property and respected this dedication. Our customers were relieved. It became another solid block of trust doing business and solidifying the oil-free community interdependency. We all understood that patents were a temporary protection, albeit a very important one, and in the long term, we had to deal with competitors and win the business by providing the best product in quality, value, innovation, and service.

The timing could not have been better, because thanks in no small part to that dedication in DTC to execute the new strategy, our major changes to the way we innovated and focused on a smaller number of customers with a clear value proposition had led to our first ever positive bottom line in 2010. We ended the year with 25 percent growth—4,943 compressors sold, with McQuay, Coffely, Smardt, Climaveneta, Geoclima, and Multistack representing about 95 percent of the volume and revenues of $71 million, but still with a modest EBITA and free cash. A mere forty-five compressors were sold in China that year, representing less than 1 percent of our sales.

Looking back, one would think that would have been a time for great celebration. Instead, there were mixed feelings: relief that Turbocor was finally profitable, embarrassment that it had taken fifteen years to make that first dollar. The overall sense was that it was better to keep moving forward with what DTC was doing in terms of streamlining processes, focusing on delivering quality products, and serving closely the few customers we were maintaining.

The work continued as DTC lengthened its reach across the world. We traveled to China twice a year and visited potential customers like Haier, Gree, Midea, McQuay, and York Zha, who had been enamored with Turbocor's oil-free compressor design a decade before. He had given up his job and poured all of his resources into starting his own company in hopes of expanding what was at the time a very fledgling oil-free community.

When we arrived at his facility, I was surprised that it was more or less an empty shell. It was York, myself, and a few other people—his entire company. He informed us that he was going to hire the most talented Chinese people from Tsinghua University and with them join the oil-free technology movement in China. He wanted to be "the new Smardt in China," referring to the successful and growing manufacturer of oil-free centrifugal chillers that was founded by Roger Richmond-Smith. York announced that the name of his company was going to be BSE, Believe in Saving Energy.

It sounded promising. Of course, promises were not the same as plans, so we continued the conversation. York's partnership with Tsinghua would allow the university to monitor technical data, which he would then showcase to the HVAC industry and convince them of oil-free technology's efficiency.

One of the key leaders with him was Professor Wei from Tsinghua University, with whom York had collaborated. He was convinced that education was the key, and the university was uniquely positioned to provide assistance to BSE and the oil-free community. York took careful, deliberate steps to build his business in the exact way he'd laid out. We were excited by the prospect of BSE partnering with Danfoss Turbocor. There were probably twenty-five small companies in China that were eager to do business with us, but none of them had the same collection of technical skills and passion for energy efficiency, thanks in no small part to BSE's connection with Tsinghua University.

The year 2012 also marked the start of implementing the "Fast Dragon Strategy." A key element of our strategy was the market data showing China as the most dominant chiller market in the world, buying one in every two chillers sold in the world at that time. Investments in China's growing infrastructure and the rising cost of energy made it clear that to be a relevant global player, we needed to win the Chinese market, and the time was now. Winning in China put pressure back on the U. S. and Europe to get one of the three global giants, Carrier, Trane, or JCI, on board with Danfoss Turbocor. It was a chess game, not a sprint run, meaning it had to be planned carefully. All understood it would take time.

We also understood that it was a more open market able to adopt new technologies and didn't have the same preconceptions and non-service business models attached as the other, mature markets did. The Fast Dragon approach was based on three key elements: win one of the top three strong local brands—Haier, Gree, or Midea; deploy a strong marketing and education campaign (pull strategy); and build a strong local team that truly believed in and understood the oil-free vision and the unconventional business model.

There were three layers of resistance to oil-free technology in the Chinese market. Traditional U. S. chiller giants (Carrier, Trane, York, etc.) dominated the chiller business and were reluctant to change their service-based revenue models. The distribution channels were conflicted, often selling both Haier and lower energy efficiency chillers at the same time. And end users were skeptical about energy savings and product stability.

This proved challenging to Haier in particular, according to Wang Li, former president of Haier air conditioning. "We had to position oil-free chillers as a high-efficiency, low-maintenance alternative, targeting end users to educate them directly rather than relying on traditional distributors," she said. "Contrary to expectations, technical staff were eager to adopt advanced technologies that improved efficiency and user experience."

Haier's initial marketing efforts, such as seminars conducted in tourist areas, had limited impact. Haier even tried to advertise on the giant screen in New York City's Times Square. Success came from real-world demonstration projects with sensors, monitoring, and data transparency. "Building trust through verified energy savings and chiller retrofit comparisons was key to shifting perceptions," Wang explained.

Mr. Yu Xinglai, general manager of the property equipment department of Shenzhen Merchants Real Estate Co., Ltd, recalled that his organization was the first enterprise that dared to try something new. "Since the Haier oil-free chiller was put into use, it has operated stably for eighteen years," he said. "The overall property is more than one-third more energy-saving than similar buildings, and the chiller is more than 52 percent more energy-saving than traditional lubricated chillers. It has obtained the national green building three-star design label certification and was selected by the Ministry of Construction and the National Development and Reform Commission as one of the thirty-five energy-saving demonstration projects in China."

This growth of magnetic bearing compressor technology had advanced rapidly over a six-year span. Professor Wei from Tsinghua University tracked that evolution of building energy conservation efforts:

- In 2005, Tsinghua University established the Building Energy Conservation Research Center, focusing on energy-saving technologies.
- In 2007 and 2008, the center participated in a demonstration project under China's eleventh Five-Year Plan, installing the first maglev compressor (made by Haier) in a real building, South China Sea Building 3 in Shenzhen.
- During an internship in 2008, Professor Wei and his students conducted field measurements, impressed by the compact and efficient design.
- In 2010, he attended the LSE Annual Awards in London, where a project with Swire Properties

won the Low Carbon Travel Award. The maglev oil-free compressor was recognized as Innovative Product of the Year.

- His connection to Danfoss Token began in 2010, through General Manager Zhong Jiang, whose business partner was a former student of Professor Wei's mother.
- In 2011, Professor Wei met Mr. Jiang in Shanghai and discussed the potential of maglev technology, especially for retrofit projects due to its flexibility and efficiency.

In addition, the Xijiao Hotel in Shanghai became a flagship project, demonstrating the compressor's reliability and performance. As time went on, the technology was adopted in hospitals, hotels, and even Taikoo Li (a massive commercial property in Beijing built before the Summer Olympic Games in 2008). Real-world data showed energy savings of 64.7 percent compared to traditional screw compressors.

The foundation for Danfoss Turbocor's expansion was there. To start building a strong local team, we needed to find someone with no HVAC background, an entrepreneurial spirit, and solid technical and marketing knowledge. After an extensive search, George Li joined as the leader for Danfoss Turbocor in China.

Li was raised in China before completing his higher education in Sweden. His worldwide education gained him fluency in three languages and positioned him as the general manager, and later commercial director, of a mobile phone infrastructure company in China. This experience enabled him to understand both Chinese and Western cultures, which gave him both advantages and disadvantages as, depending on whom you talked to, he was either an insider or an outsider. Another key factor was that we were bringing someone from outside the HVAC industry to provide the focus, speed, and autonomy we need in China.

"It was two sides of the same coin," Li recalled. "My background was mostly outside the HVAC industry. I'm an engineering physicist by training, so I'm not unfamiliar with engineering and concepts, but I had a learning curve in HVAC. At the same time, I didn't have the burden of tradition, meaning I had a different view of what was possible and what was not possible."

Li was impressed by the DTC business model and the oil-free compressor technology and immediately worked with us to support the expansion of that vision to the Chinese market. He had no illusions about the difficulty of doing so, but like so many other visionaries who have joined Turbocor throughout the years, people who often come from outside the norms of the compressor field, he was ready to take a chance.

"We want to do something that might be hard, understanding the hardships, but we believe we can do it and reshape the HVAC industry in China," he said. "We can do the right thing, because we believe in ourselves. If we encounter trouble, we solve the problem. We don't blame each other. I want to spend the next ten or fifteen years with our people as opposed to just a brand name on a door."

Perhaps the greatest obstacle that Li and others faced in the Chinese market was the skepticism about how efficient the Turbocor compressor really was. "How can this be possible?" said Li, quoting potential customers, suppliers, and partners. "'How can this be so efficient? And also, how can I trust you to be here in this country, long term, so that we can build a long-term relationship around manufacturing and end users?' So there's a lot of trust building involved in the beginning. And that's probably the biggest challenge, that we kind of went in there as an innovator, a disruptor even, and the first step, aside from introducing ourselves, was to gain trust." As Ms. Wang and Professor Wei mentioned, we all understood the challenges to educate the industry.

Li became such an ardent proponent and advocate of the magnetic bearing technology that he was soon able to begin forging those connections and building trust with customers and end users. To add further support, we sent Heath Whiddon and his family to Shanghai. Whiddon, a former service engineer from Trane who joined us in 2008, was our best service and HVAC expert and a true believer in oil-free technology. He trained our young organization and new customers with knowledge and passion for the technology that was contagious and critical in those early days in China.

"Going to China in 2013 was a life-changing experience for me and my family," Whiddon later said. "It wasn't just about work; it was about building something meaningful. I helped set up our sales and service office, trained local teams, and shared everything I knew about Turbocor compressors."

Whiddon pointed out that China was the largest chiller market, and DTC saw the potential early. "The OEMs were cautious at first, but once they saw the value, momentum grew," he explained. "Today, the team there is strong and self-sufficient. I'm proud to have played a part in that journey."

It was the same trust that DTC was establishing with companies in the U. S. and Europe and that it had tried to create during the previous years of organizational and economic crisis. The oil-free community started to take shape in China on the same push and pull strategy implemented in North America and Europe.

During the 2011 to 2012 era, after we stopped selling to the small retrofitters and began focusing on a few key clients, Turbocor found itself enmeshed with a handful of OEMs who, together, were responsible for 80 percent of our revenue: Daikin McQuay, Axima, Geoclima, Multistack, Climaveneta, Haier, and Smardt. Working hand in hand with these companies to master the use of our technology and better integrate it with their products meant we had to form close relationships with these few customers. Turbocor's future hinged on the personal connection—or customer intimacy, as we called it—with a handful of people in the oil-free community. A mutual dependency was created with a goal to educate end users and change the HVAC industry.

Maintaining that balance was complicated because, as we narrowed the field, some were muttering that our extreme selectiveness was arrogant. To further our goals of being choosy, we devised a questionnaire to, in effect, interview our potential customers. That questionnaire asked:

> Can you sell one hundred oil-free chillers per year?
>
> Can you sell high-efficiency chillers that are 20 to 25 percent more expensive than the traditional lubricated ones?
>
> Can you invest $15 million in research and development to master this technology, energy efficiency sales skills, and building a service organization?
>
> Can you invest in a laboratory to demonstrate the chillers' efficiency?
>
> Can you train your sales force to master how to sell total life cycle cost based on energy efficiency?
>
> Do you have the capital and competence to build a strong service network?

Many soon found that if they couldn't do all of these things, they weren't going to be our customers. Some were upset at first. How dare we turn down their business? Wasn't their money as good as anyone else's?

Still others came to us later and said, "Thank you. Thank you for making us go through this process, because we didn't realize this was something other than a normal compressor, just another mechanical device we could plug into our chiller. We would have lost money on those developments, and we would have been unsuccessful."

We encouraged these customers to discuss with our OEM partners and evaluate a potential license or distribution agreement for the oil-free chillers with Danfoss Turbocor compressors. Multistack and Smardt ended up doing so and extended our oil-free community to other small regional OEMs.

We understood that with few and relatively small OEMs, our push strategy would limit our potential, as the industry needed to be re-educated to understand the total life cycle cost of the chillers, not the initial price or full load efficiency on their own. DTC had to complement its approach to the market with a strong pull strategy to create the demand. Convincing the conservative industry about the benefits of the oil-free compressor technology meant using case studies to demonstrate the hard data for our strong value proposition. The combination of the push and pull strategy was a strong part of our journey moving forward to change the status quo with our selected OEMs. We would keep developing the technology with innovation and focusing on it to make it a great business.

Perhaps a greater step was the hiring of a firm to contribute to two key internal documents: the *Comprehensive Plan for Profitable Growth* and the *Cost Reduction and Detailed Action Plan*. We asked for some assistance, but Danfoss could not provide the needed expertise to our specific requests for the material cost reduction initiative, so we looked externally.

We contacted a consulting firm, recommended by Kim Fausing, that had deep experience in the automotive industry in Munich, Germany. Why automotive? Because the power electronics and high-performance electrical permanent magnetic motors we were developing were similar to what advances were taking place in the early years of electric vehicles.

The gentleman leading their team, Marcus, came to Tallahassee as we whittled down our goals for DTC and its continued growth. After reviewing our data and answering some tough questions, we reconfirmed our initial priorities on two applications, water-cooled and air-cooled chillers from 100 to 400 tons of cooling capacity, and selecting new markets like China, the Asia-Pacific, and the traditional markets in North America and Western Europe, while de-prioritizing South America, the Middle East, and Eastern European countries.

The consensus was clear: we were on the right path. We could not be all over the map. We had to focus on very specific customer applications and markets.

The greatest benefit of this exercise was that it forced us to take a closer look at our material cost reduction so that Turbocor could continue to operate in the black, improving our contribution margin. As noted before, the approximately 20 percent increase in our pricing had rectified most of that price issue. The question then became how to cut back on materials cost in our production. The timing was better than we could have planned for, because we were about to redesign the product completely in order to create the next generation of our compressors at a lower cost. But Marcus said, "Instead, spend your time concentrating on your existing platform. Look, here are the huge opportunities that you have."

Brainstorming with an automotive boutique consulting firm led us to do a deep dive on cost reduction opportunities, like changing the design from large to small magnets and standardizing power electronics, resulting in overall cost reductions between 20 and 30 percent. This was a huge win for Turbocor, and it could not have arrived at a better time.

The days of the joint venture were coming to a close. Danfoss had watched Turbocor's development in the past decade as the company struggled, overcame, and grew. Our name was spreading across the world. Our oil-free compressor technology was drawing a devoted following of believers in greener HVAC solutions. Our comprehensive five-year plan impressed the Danfoss board, and by 2013, Danfoss was ready to acquire Turbocor fully, completing a venture that had begun in 2004.

"Our goal from the beginning was to acquire Turbocor 100 percent," Kim Fausing said years later. "That was not doable in the early years. I think it was really about creating a safe environment and making sure we were building on the positive notes for the future."

I was in Brazil during the new year break when the call came through at the beginning of 2013. It was a tremendous relief once Turbocor became a full member of the Danfoss family. It is significantly easier to have a single owner, given the complexity of a fifty-fifty joint venture, and Danfoss, with no doubt, was the best. Our customers were pleased by this show of good faith, that Danfoss was investing in Turbocor's outcome.

With this exciting change came additional expected challenges in the shape of asserting oversight from Danfoss. This included ways to improve manufacturing productivity, standardizing processes, and reducing waste. This program was recently implemented by the Danfoss Business System (DBS) across all businesses and was extremely successful, and Danfoss Productivity Program (DPP) was delivering 10 percent year to year labor productivity improvement across all the mature products in manufacturing. The orchestra was becoming bigger and better arranged through this standardization.

To that end, a few weeks after the full acquisition, we started to be overwhelmed with people from our headquarters coming to Tallahassee with the objective of providing support, including a gentleman from the Danfoss Business System. That assistance was focused solely on labor productivity, and it became clear after touring the manufacturing floor that cutting positions was necessary—30 percent of the manufacturing workforce.

We were adamant that no personnel cuts take place. Labor costs were less than 2 percent. We planned to address productivity by instead improving material cost reduction, quality, and finishing product development as spelled out in our strategic plan. One reason we had a higher number of people in shaft production than might be standard was because we had an extremally complex product, and quality issues and product development were going on in parallel to standard compressor production, so people were there to fix problems that showed up on the line and support prototypes.

Eventually, this disagreement led to his departure, and while no one from higher up in the organization called to overrule us, the pressure remained for DTC to conform to considerably rigid structures that had proved successful for the rest of Danfoss. I was convinced, however, that they didn't yet fully understand what was happening at Turbocor and how it needed to maintain some independence. But if anyone would understand, it would be Kim Fausing, the man who had served on the joint venture board for several years up to that point and who had come up with these structures. I was convinced he would advise us how to set a winning strategy.

I was faced with a difficult decision that could have a drastic impact on Danfoss Turbocor and potentially the customers' relationships, let alone the direction of the company. Danfoss and DTC would have to find ways to work together if this nexus of the oil-free industry was going to survive.

Scan the QR code below to learn more and view pictures.

# Part Four—2013 to 2018: Customer Intimacy and Founders Mentality

CHAPTER EIGHT

# The Pivot East: From Florida to Haiyan and Back Again

The visit to Tallahassee by Danfoss representatives left us unsettled. They came with good intentions, like their desire for greater productivity, but their presence stirred confusion. We were caught between preserving our independence and embracing the promise of being fully part of Danfoss.

As 2014 dawned, we faced the distinct possibility that we would no longer be able to preserve the company's entrepreneurial, innovative, and visionary spirit while hewing to a "One Danfoss, One Way" directive that was challenging the flexibility of our organization. The "Core and Clear" strategy introduced by Danfoss in 2010 focused on getting the basics right, standardizing the process and procedures of our company on a global scale.

The acquisition in 2013 dissolved the Turbocor board, simplifying ownership but removing the Australian contingent from management. We were integrated into the Commercial Compressor (CC) business, and I began reporting to President Leif Flojgaard. Leif was a respected colleague, but he struggled to see the differences between Turbocor's disruptive model and the traditional scroll compressor business. Reconciling those worlds was not easy.

Meanwhile, we worked to protect our way of doing business. I quietly documented the corporate thrust toward standardization and the risks it posed. Some initiatives in finance, HR, and IT were beneficial, but others, including the recommendation to eliminate the Turbocor brand, were alarming. Customers weren't asking for "a Danfoss compressor." They wanted Turbocor technology. The brand carried years of innovation, awards, and credibility. Losing it would mean losing trust.

The biggest challenge came when Danfoss sought to integrate our sales, application engineering, and customer service into regional structures. This would distance us from our loyal customers and shift us from our push and pull strategy to a transactional model. Concerns mounted, and the loss of our VP of sales and marketing, Doug Bishop, further increased the risk. For the leadership team, it felt like déjà vu from 2004, when the joint venture had unsettled both employees and customers.

I drafted a detailed report for the CC president, outlining how these moves could destabilize Turbocor and put the acquisition at risk. I included recommendations from the leadership team and, reluctantly, the possibility of my resignation. It was a painful step after so many years with Danfoss, but I felt compelled to raise awareness. I shared the report with Lin Sun, and together we committed to protecting our people and customers.

Before sending it, I sought advice from my mentor, David Loveless, a retired CEO and Harvard alumnus. He urged me to reach out directly to COO Kim Fausing instead of resigning. His counsel proved pivotal. The next morning, I contacted Kim, who quickly scheduled time with me. He listened quietly as I explained the threats to our priorities: quality, execution, cost reduction, and customer intimacy. I asked for support to preserve agility and independence.

Kim acted discreetly so we could maintain some measure of autonomy. He allowed Turbocor to operate as an insurgent within Danfoss, guided by three principles:

1. Insurgent Mission: Drive our bold vision to transform HVAC with oil-free technology.
2. Frontline Passion: Preserve intimacy with our select oil-free partners.
3. Owner's Mindset: Run the business with responsibility, urgency, and long-term commitment.

The same principles were later well documented by Chris Zook and James Allen in the book *The Founders' Mentality*, published in 2016 and presented in the Danfoss GMM in 2018 in Italy. This was significant because Danfoss was showing what it looked like to balance and differentiate the insurgent and incumbent business in Danfoss.

*Danfoss Business System*, which was our playbook for all entities, brought tremendous results to the Danfoss group. Everything had to be standardized. There had to be one template. Each manufacturing site and each office had to adhere to a new form of internalized discipline. It was absolutely needed for a mature and successful business like Danfoss. All of this was done in a way that the Danfoss Turbocor leadership team and organization was not involved, protecting them and Danfoss from similar hard feelings experienced in the early days of the joint venture.

As additional issues were resolved, Noel Ryan became my new boss. Based in France, he understood compressors well and visited Tallahassee early on. He challenged our budgeting process annually, ensuring accountability. Our bottom-up planning system, created by CFO Stan and mastered by Sha Ouyang, encouraged managers to run departments like their own businesses. It was demanding but fair, and it kept us disciplined.

From 2011 to 2013, revenue held steady between $71 and $75 million, but EBITA and cash flow improved significantly. This reflected our focus on a few customers, educating the industry on efficiency and life cycle costs while tightening quality and cost. By 2013, our Compass emphasized improving quality by 30 percent, reducing costs by 10 percent, and strengthening financials, thus reining in unfettered innovation to secure profitability.

As profits improved and new customers came on board, we brought in consultants from Palo Alto led by Greg Madsen to reignite innovation. The culture was still alive, but dormant. It was time to wake the giant. Madsen warned that without regular new products and solutions, even pioneers risk becoming average. That was our fear: that Turbocor had swung too far from bold creativity into conventional, slow execution. Customers depended on us to expand oil-free chillers into air-cooled and water-cooled applications.

Madsen urged us to embrace five behaviors:

*Associating.* Connecting seemingly unrelated ideas to spark new insights. Example: Steve Jobs combined calligraphy with computer science to influence the design of Apple's typography and user interface.

*Questioning.* Keep asking questions to learn, like a five-year-old child. Why that? How does it work? Ask provocative questions that challenge the status quo. Example: Jeff Bezos asked, "Why can't people buy books online?"—a question that led to the creation of Amazon.

*Observing.* Watching people, macro trends, processes, and environments to gain insights. Example: Howard Schultz observed Italian coffee culture and used it to shape Starbucks's customer experience.

*Networking.* Interacting with people from diverse backgrounds to gain new perspectives, networking with people in person or virtually outside your comfort zone in aerospace, electric vehicle manufacturing, and university education.

Example: Innovators often attend conferences outside their field to gather fresh ideas and challenge their assumptions.

*Experimenting.* Actively testing new ideas through rapid prototyping. If you have to fail, do it fast, without spending too much money or exposing your brand and reputation. Example: Thomas Edison famously tested thousands of materials before finding the right filament for the light bulb.

This became our playbook. We invested in R & D labs, prototyping workshops, and new talent. We built processes to manage both radical and incremental innovation, aiming to launch one new product or service each year. In spring 2015, Madsen took our leadership team to Intuit, Tesla, and Google. At Intuit, we saw how innovation practices could attract talent. Tesla showed us their focus on cost versus our efficiency-driven design. Google revealed the power of spare time and rapid prototyping.

Coupled with Christensen's philosophy and *The Innovator's DNA*, these lessons reinforced the balance between innovation and execution. Turbocor had long swung between extremes, but now we were combining the rigor of business discipline with entrepreneurial risk-taking, alongside customers who trusted us to keep innovating.

In 2015, we restructured sales to strengthen intimacy and the push and pull strategy. Four regions—the Americas, Europe, China, and the Asia-Pacific—reported directly to Turbocor leadership. Jose Alvares led the Americas, George Li led China, David Williamson moved into sales for Europe, and Scott Morehouse continued his strong work in the Asia-

Pacific. This structure, paired with renewed innovation, drove a 70-percent growth from 2014 to 2017, crossing $109 million in revenue and doubling the business since acquisition.

Recognition followed that same year. We won the Innovation Product of the Year at the AHR Show in Chicago for the VTT Series Compressor and the Alliance to Save Energy's Innovation Star of Energy Efficiency Award. The gala celebration, shared with partners like Angelo Troiano of Arctic Chill, Charlie Fletcher of EMCOR-MESA, Vince Canino of Smardt, and Lisa Tryson, marked not just awards but validation of Turbocor's disruptive vision.

Expansion into China, despite humble beginnings, had become a significant part of furthering that vision. In 2010, we opened a small service center in an empty office space. Three men sat at three tables, repairing compressors day after day. It was a good start, because our people there were able to put their technical skills to good use fixing issues that took place with these high-end pieces of machinery. We took pride in those modest surroundings, because our people were solving problems for customers who trusted us, and it marked our first step toward something bigger.

Meanwhile, our Danfoss colleagues were busy selling over a dozen different product lines to thousands of customers, including small valves, controllers, scroll compressors, and heat exchangers. They were selling a huge basket filled with many different mature products through the conventional-channel OEMs and distributor-typical push strategy. Along came Turbocor with its expensive, advanced magnetic bearing compressor, which could not quite keep up with the cheaper devices in terms of purchase price but could, in the long run, compete well in terms of energy efficiency. This was extremely appealing, given how the cost of electricity in China was rising during those years and the speed at which the country was developing.

Lin Sun and I spent considerable time in China, supporting George's small team and working closely with our customers. It's no exaggeration to say that Lin played a pivotal role in our growth and success in the Chinese market. As the Danfoss Turbocor CTO who was instrumental in developing and designing the Turbocor product, Lin brought both technical expertise and powerful leadership that resonated strongly in China. He was seen as the "Chinese brain" behind Turbocor, an engineer born and raised in China with deep ties to Tsinghua University. He carried credibility that none of us could replicate . We admired how his cultural fluency and personal warmth opened doors that would have remained closed otherwise.

Our CFO, Stan Zalik, was initially skeptical about building a factory in China. He worried about price erosion and patent infringement. We traveled together for two weeks, meeting customers and officials, hopeful that learning about the emerging market would change his mind. He came back as our biggest supporter for that expansion because he understood it as a huge growth opportunity and saw we were taking the appropriate actions to mitigate the risks. He recognized that we should put a manufacturing site in China as soon as possible.

At the time, we were making small, steady progress in China, but it was not until 2013 that a delegation of Haier executives, including Ms. Wang, Mr. Fu, Henrik, and Mr. Guo, visited our Tallahassee home base, allowing us to discuss face-to-face for a few intensive days what we should do together to bring oil-free compressor technology into the mainstream in China.

Haier, headquartered in Qingdao, China, was a global manufacturer of household and commercial appliances with the ambition to be a global leader in commercial air conditioning. Wang Li remembered her first encounter with magnetic bearing technology while working in Haier's kitchen division in 2007 to 2008. Though not directly involved in commercial air conditioning, she was aware of internal efforts to develop a leading position and not follow the traditional global players and technologies. The emphasis on high energy efficiency came from Haier's leadership, particularly Mr. Liang, who aligned with CEO Zhang's broader philosophy. The goal was to position Haier as a leader, not a follower, in the chiller business.

This realization was no coincidence but was based in part on the increasing electricity costs and air pollution coming from burning fossil fuels to generate electricity in the fast-growing country. The result of that visit to Turbocor was that Haier was convinced that oil-free compressor technology was the best way to position themselves as leaders in China who were dedicated to handling those costs and preserving the environment.

Discussions revolved around marketing and education, and manufacturing and supply chain, an especially important focus because Haier had started up a chiller business in the corner of one of their factories in Qingdao. We noted that they would have to make significant investments—in research, manufacturing, and development, of course, but primarily in educating the market about energy efficiency and total life cycle cost of operation of chillers.

Haier took this as a friendly challenge and, in rapid time, was able to design and build a brand-new, state-of-the-art manufacturing center that would make chillers. We went to see it in 2015, before it opened officially in 2016, and were absolutely astounded at the huge, sprawling, modern space, the clean lines and bright lights, and the way our voices were lost in the cavernous ceiling as machinery and robots rumbled all around us.

It was our turn, then, to bring our manufacturing to China. Despite skepticism about timing and volumes, we decided to act. With strong cash reserves, we invested $4 million in a U. S. production line, then shipped it piece by piece to Haiyan. We remember the mix of anxiety and excitement as we took that risk. Our philosophy was simple: better to ask forgiveness than permission.

By 2017, under Paul Dean in Tallahassee and Ping Chen in China, the facility was running. Compressors rolled off the line after eighteen months, far faster than the conventional three years. We felt immense pride in our team's determination and agility.

The grand opening in April 2017 was a banner day for Danfoss Turbocor and the oil-free community, with a guest list that brought the bulk of the oil-free community together, including Kim Fausing, Jesper Christiansen, Ron Conry, Roger Richmond-Smith from Smardt, Mr. Li Hui Jun from McQuay, York Zha from BSE, and Wang Li and Mr. Guo from Haier, along with many others from around the world, in addition to the core of the oil-free community and the Danfoss Turbocor team.

It was quite remarkable to chat with people who, at one point or another in the previous twenty years, played a role that wasn't limited to developing the magnetic bearing compressor technology, but also encompassed making sure the company that offered it could survive—entrepreneurs and visionaries, all. We found ourselves reflecting on the journey from three tables in a bare office to a thriving factory in China. Looking around at the entrepreneurs and visionaries who had shaped magnetic bearing technology over two decades, we felt both humbled and inspired. What was meant on the surface as the celebration of a new plant became proof that persistence, risk-taking, and belief in innovation would carry our vision and our product across continents.

The establishment of this factory kicked off a new era in oil-free compressor manufacturing. In 2014, Turbocor sold 397 compressors in China, and all of these were produced in the U. S. and accounted for a small 7 percent of our global volume. A decade later, by 2024, China had become our largest market, accounting for a staggering 37 percent of global sales, with nearly six thousand compressors produced locally.

The growth was staggering—fifteen times larger—but not everyone believed in it at the start. Frank Ford, then director of product management, admitted he was skeptical. "From an economic standpoint, the overhead of supporting two sites seemed too much," he recalled. "But it turned out to be one of the most critical decisions we ever made."

That success was built on the skill of key people in China. Ping Chen, a rising talent in Danfoss, was recommended by Arthur Xu to lead the new facility. She hired local experts and organized what began as an empty shell, piecing together the dismantled production line shipped from the U. S. It became our first attempt at higher automation and tighter quality control, a lesson that shaped the new Tallahassee factory years later. Meanwhile, we synchronized operations between Haiyan and Tallahassee: if China produced ten units a day, we shipped ten kits of shafts, bearings, and sensors to match.

The timing proved perfect. For years, we had tried to secure Trane, one of the main OEMs, as a customer. Deals nearly closed three times but fell through, largely because Trane's U. S. culture insisted compressors had to be designed and built internally. By 2016, however, our growth caught the attention of Jose La Loggia, president of Trane Europe. With Argentinian and Australian roots, he had seen Turbocor's dominance in Australia, its growing position in Europe, and especially in China. His view was pragmatic: if together we could grow high-efficiency market share and deliver life cycle savings, why should it matter who produced the compressor?

Carlos Zamudio, our senior Trane global accounts manager, knew it would be a tough sell. "Trane was buying internally," he said, "but La Loggia realized he wasn't getting the innovation he needed from screw compressors in Europe." He leaned on Turbocor instead. Within eighteen months, our compressors were running in a Trane lab in France, destined for a hospital in Turin. When La Loggia visited, he asked to turn on the chiller. The team led by Yves Raimbault and Ray Good explained it was already running—at 40,000 RPM. Shocked by the silence and smoothness compared to traditional screw compressors, he became a strong advocate of oil-free technology.

The market was stunned. Trane had finally embraced what Turbocor had been championing for years. Even in the U. S., people began to reconsider. Zamudio reflected, "This is the most complex machine we sell in Danfoss. Customer intimacy is critical. What Kim Fausing afforded Turbocor was the ability not to compromise, even when "One Danfoss, One Way" was the mandate. That's why we have today's lineup with the TGS 380 and TT490 and why we're succeeding in data centers."

It was around the same time that Turbocor was cementing its position not merely as the lead in oil-free technology in the United States and Europe, but also as a place where we could study the efficiency of this advanced technology. To that end, we wanted to open an application development center modeled on the European ones Danfoss used to bring customers in to, in a sense, get their hands dirty.

The hydraulic business division had dreamed these up to bring in customers. There they could change components on tractors, test-drive the vehicles, and put their heads together to experiment with ways to improve solutions to hydraulics issues. After a while, the idea trickled to people working in climate solutions that a similar concept should come to the U. S. to collaborate with the refrigeration and air conditioning customers.

The center would serve innovation and was customer centric to all Danfoss products, yet it would be a great move for encouraging collaboration between different divisions of the company, surrounded by world-class labs and research at MagLab, FSU, and FAMU.

Construction ran through 2016 and 2017, alongside the Haiyan factory. When the center opened, customers were astonished. Walking through glass doors into a gleaming white building with Danfoss's signature red walls, they could see compressors being built, tested, and applied. They left inspired by the innovation they witnessed.

Today, the application development center continues to expand, from residential air conditioning, rooftop units, and commercial refrigeration to being able to test an entire air-cooled chiller of up to 150 tons. That section of the building is bigger than a house and is equipped with sensitive climate controls so that technicians can run the chiller at the minus fifteen degrees Celsius it might encounter in Minnesota or at the fifty-five degrees it could experience in Arizona. Customers who are actually buying compressors from Turbocor can use this lab for free to test the endurance and efficiency of their chillers. It's secure, too, so much so that visiting representatives from other customers cannot go inside if a test is underway for one of their potential competitors.

The groundbreaking day remains vivid. Mads-Peter Clausen, grandson of Danfoss's founder, joined us with representatives from Denmark. Together, we dug the ceremonial first shovelfuls of dirt. Mads-Peter used a special shovel once wielded by founder Mads Clausen at factory openings worldwide. It was a powerful symbol that ushered Turbocor into a new era of expansion and success, turning the spotlight on oil-free compressor technology as deserved.

Scan the QR code below to learn more and view pictures.

CHAPTER NINE

# Founders at Heart: Keeping the Startup Spirit Alive

A considerable amount of focus thus far has been on Mads Clausen's vision for highly efficient cooling systems and how his son, Jørgen, carried that vision into the twenty-first century, leading first to the acquisition and then to the expansion of Danfoss Turbocor. It's equally important to see what the next generation of Clausens has achieved in the continuance of that legacy.

Jørgen's brother, Peter, was involved with Danfoss for a time but later removed himself and stayed on as a member of the company's board and the Danfoss Foundation. Peter's child, Mads-Peter, decided his interests would lead him back to that engineering bent that made the family's innovation and entrepreneurship vital.

Mads-Peter Clausen first whet his appetite for the HVAC industry, in which Danfoss had been working for so many years, by helping us put a scroll compressor factory in

Tianjin, China, during the company's early days. Shortly thereafter, during the start of our air conditioning journey heading up global accounts in Baltimore, Kim Fausing called me one day about bringing Mads-Peter into that branch of the business. I readily agreed on the condition he be treated like any other member of our team.

All of us, especially Mads-Peter, were happy with the condition. We bonded well in our appreciation for the Danfoss brand and the goal of pursuing energy-efficient and clean cooling machinery. He worked in our division for several years and stayed in Baltimore after I moved to Tallahassee to take on the leadership of DTC, before leaving to pursue his MBA at the University of Georgia in Atlanta.

One of the next times I saw him was not long after I had the critical call with Kim regarding Danfoss Turbocor's independence following the 2013 full acquisition. Mads-Peter came to Danfoss Turbocor for a casual visit as he was leaving Atlanta, adding a quick four-hour drive to Tallahassee, and joined Juliana and me for a dinner at home. It was a nice time of catching up, both personally and professionally. He asked me my opinions about what was going on with Turbocor. I didn't shy away from the truth any more than I had done during the call with Kim a few months earlier, since we had been teammates for so many years.

It was clear that the call with Kim had reached the family leadership and reignited the "founder's mentality," the entrepreneurial spirit that had long driven Danfoss's DNA. That moment reaffirmed and encouraged our commitment to pursuing the bold vision of transforming the HVAC industry with our high efficiency oil-free platform.

A few years later, in the summer of 2016, Mads-Peter came to work for us in a similar arrangement to what we had in Baltimore, being treated as another member of the Danfoss Turbocor team. He had a small apartment in downtown Tallahassee and lived simply, although he could have easily thrown around his weight, so to speak, as the grandson of Danfoss's founder. Instead, he continued his quiet dedication to the job, as he had always done. He'd show up parking his Ford Taurus among the spaces used by all of us DTC employees.

Few people knew that he was one of the biggest shareholders of the company. Of course, it was obvious there was a new Danish guy with the last name Clausen joining our Tallahassee team, but if they connected the dots, they were not treating him any differently.

His work with the oil-free development technology was invaluable. Part of his work focused on developing the Danfoss complementary oil-free portfolio, as we were lagging behind in valves and controls suitable for operation without lubricants. We were grappling with problems regarding those valves, because it turned out that when we eliminated the oil from the system, the valves that used to have that oil as a lubricant began to malfunction. Mads-Peter's goal was to go beyond assisting us in finding solutions. He wanted to make sure Danfoss could provide all manner of parts that went along with the compressors rather than the component by itself.

It was an astonishing experience to have one of the Danfoss shareholders working in our organization. He was incredibly humble and dedicated to daily projects while leading a small team. This sent a strong and clear message to all of us—the Danfoss Turbocor team, Danfoss, and our customers—about how deeply rooted the Danfoss values of hard work, humility, and teamwork were, regardless of one's position within the organization.

When he did eventually finish up with us by the end of 2018 and move back to Denmark, we were all sad to see him go, but we all knew it was a temporary assignment, and Danfoss Turbocor had been honored to have him. We felt we had converted him to a true believer in oil-free technology and a great advocate of energy efficiency. We were happy when, shortly after, it was announced that his father, Peter Clausen, was retiring from the Danfoss board and Mads-Peter would be his replacement, both there and on the Danfoss Foundation board. At the same time, Jørgen Clausen was retiring as a Chairman of the Danfoss board and Foundation board member, with his son Mads Clausen taking a board member position at the Danfoss board and Foundation.

All of those things serve to demonstrate how much of a family enterprise Danfoss remains after ninety-three years, carefully working to maintain the core values that direct Danfoss into the future with the next generation. This is why we were delighted to have Mads-Peter in Tallahassee for those short two years and to see him take up the shovel at the groundbreaking ceremony for our application development center. He carried the founder's name and the founder's spirit, demonstrating to DTC employees that the entire Danfoss family was valued not for their lineage but for their contributions.

It was that value that consistently lifted our team to greater heights in terms of experimentation, a constant desire to innovate that resulted in our TTH compressor's launch in 2018. This high-lift model opened up an entirely new front on which Danfoss could approach the global HVAC market.

Traditionally, centrifugal compressors run with a low-pressure ratio. This makes them suitable for specific applications and highly efficient when typically elevating pressure from 1 to 4. However, the TTH was engineered to test those boundaries, achieving a pressure ratio from 1 to 6.2, a level previously unseen in the HVAC industry in centrifugal compressors air-cooled application. It came out of an older project that had been toyed with, one that would create a three-stage compressor instead of our typical two-stage.

That concept was too expensive because it would be complex. That complexity arose from the need to balance the whole shaft with three impellers, as forces are pushing in one direction, making it a potential need to change the design of the magnetic bearings.

Instead, our engineers created a shaft with only two impellers that were back-to-back, neutralizing the forces at play and balancing them among the magnetic fields. This was the result of first cutting down on the uncontrolled innovation and then ramping it back up to a level at which we could truly experiment, but under controlled circumstances that brought things like cost-effectiveness to the forefront and with the commonsense considerations that Lin Sun always brought to the picture.

To keep the compressor competitive, they designed the new product using as many common parts from the other platforms as possible, achieving about an 80-percent ratio, with what I consider brilliant engineering execution and an innovation approach.

This TTH compressor, running with higher pressure, offered a significantly more efficient alternative compared to conventional screw compressors, especially in air-cooled applications where centrifugal compressors benefit when the condensing temperature / pressure decreases. Such a compressor could be used in regions of high ambient temperatures, anywhere from the American Southwest to the Middle East. When that temperature is higher, the compressor has to work harder to bring gas from one level to another. Therefore, a higher-pressure compressor can get the job done much better.

The high pressure meant the compressor could theoretically be used in heat pump and heat recovering applications, allowing Turbocor to begin looking at systems that could generate hot water for homes and other buildings in cold European winters and in California, where the restrictions on natural gas started to lead the decarbonization discussions. A heat pump is basically a unit running in reverse, instead of producing cold, to generate the needed heat. When it's cold out, it extracts whatever heat there is outside in frigid temperatures to heat the interior. It can do this at about one-fifth the cost of fossil fuel heating methods.

This technology has been mastered by one of our early adopters, Axima-Coffely-ENGIE, initially with heat recovery with our TT and TG platforms and now with TTH and TGH rapidly expanding usage in Europe, where nations have come to realize that overdependence on natural gas from other countries could be dangerous. The initial Russian conflict with Ukraine in 2014, coupled with the expansion of the war in 2022, demonstrated that dependence as natural gas supplies were tightened.

In the U. S., the California Energy Commission started drafting new standards that would limit the usage of natural gas, to be effective by 2026. Our agile and highly innovative oil-free partner, Multistack, started to lead the way with oil-free HP and heat recovery. With the TTH and TGH compressors, Turbocor began entering the high-pressure domain, representing the first steps toward replacing traditional boilers and reducing that reliance. It was an important step in the electrification of heating.

The TTH development was a natural outgrowth of our aggressive schedule established five years before, when Turbocor decided we wanted to have a new product showcased at all the shows in China, Europe, and the U. S. Most times, those products weren't finished, because of the sheer complexity that accelerated pace necessitated, but we could, together with our close customers, at least indicate to the industry where the oil-free community was going and demonstrate the benefit of sharing risks together.

This usage is still in its early phases even today. For it to be fully useful in large heat pumps that are under development, we must raise the upper temperature of the TTH, from sixty-five degrees Celsius to eighty, and it comes with active cooling, another brilliant innovation. This will be a huge expansion as the market grows and we work toward a greater amount of dedicated compressors for heat pumps. We have no doubt that as the first wave of exponential growth for DTC came from our compressors being used in data center chillers, we'll see a second wave in heat pump development.

Why such a push? Why not take our time, two or three years, to fully develop a product for the market? We wanted to show the market we were still innovating, working side by side with our partners, advancing oil-free technology and looking for new applications, new things we could do with magnetic bearing compressors.

"Turbocor" continued to be on industry lips, thanks in no small part to the TTH and TGH compressors. As 2018 dawned, the new device captured not one or two but three awards across the world. In North America, AHR Expo selected the TTH as its Product of the Year and recognized it with the expo's Innovation Award in the Green Building Category.

Then came the 2018 China Refrigeration Expo, where it was honored with the China Refrigeration Association Innovation Product Award. This award is one of the most respected accolades in the Chinese HVAC and refrigeration industry, celebrating technologies that push boundaries in energy efficiency, sustainability, and engineering excellence. DTC also walked away with the award for innovation in sustainable HVAC solutions at the Mostra Convegno Expocomfort in Italy.

These awards touted what we and our customers had long known: that DTC was at the cutting edge of clean, sustainable cooling and heating enterprises. The Chinese award specifically praised the TTH for these qualities we had been building into our products: oil-free magnetic bearing technology, which reduced maintenance and environmental impact; high-lift capability, enabling operation in extreme ambient conditions and heat recovery systems; and low noise and vibration, which made it ideal for commercial buildings and sensitive environments.

I cannot recollect any other product in the history of the HVAC industry that received all of these awards. It was telling us, our customers, and our competitors that Danfoss Turbocor was leading the way and would continue to do so.

The awards were an important component of our push and pull strategy to educate the industry on how we could keep moving the boundaries of energy efficiency higher and higher, and this recognition coming from independent organizations across the globe gave Danfoss Turbocor and our partners tremendous credibility.

These awards were also vital when it came to reminding Danfoss about the importance of maintaining our close ties to our customers, the "customer obsession" well described in *The Founder's Mentality*. As our business in China began to grow rapidly and business from our local factory boomed, it became clear that we were entering a phase of exponential expansion in that country. This growth, and now with a factory in China, triggered a new wave of efforts to integrate our sales and marketing teams there.

However, the Danfoss sales organization sought to take direct control of customer relationships, expand Turbocor sales to a wide range of new OEMs, and define pricing policies independently. While that made sense on the surface for mature products and conventional business models, this approach was in direct conflict with our original push and pull strategy and "customer intimacy," which emphasized centralized control at Danfoss Turborcor and strategic alignment with our few strategic partners.

We resisted with the firm reminder that, as Roger Richmond-Smith put it, we were selling energy efficiency, not components. Let's say a chiller with an oil-free compressor is 20 or 25 percent more expensive than traditional, oil-using compressors on the market. That will cost a greater amount up front, of course, but the buyer will earn that difference back in a short two years if the cost of electricity is above fifteen cents per kilowatt-hour. For a market like China, where, as we mentioned, energy costs were rising to exactly that level, the key had to be to emphasize the longer-term savings and efficiency of cost, as well as environmental benefits, not the initial price.

It seems simple when explained that way, but the sales team found it challenging to alter their mindset, a mindset that was a bit transactional. They were used to selling components in the competitive market, components that would work well and be competitive, period.

When it comes to energy efficiency, it isn't enough to make the sale and walk away, hoping the customer has the knowledge to sell that ideal. It's important to educate not only the customers, but the entire market on how to apply and appreciate high energy efficiency technology like ours.

It required another adjustment from Danfoss Turbocor Sales to keep the "customer obsession" as described in *The Founder's Mentality*. George Li would then move to the U. S. to take charge as global head of services and digital business. Jose Alvares, the talented engineer I'd met in the Turbocor labs when I took charge back in 2009 and the sales and marketing leader since 2014, relocated to Shanghai, China, as our new global vice president of sales and marketing, with three regions—the Americas, Europe, and China—reporting to him. This was all done to maintain that vital customer intimacy and the business model of push and pull there. Jose and Ping worked together well as a bridge between us and our customers, ensuring we delivered compressors on time and that our mutual business model kept winning.

We remained committed to our stringent requirements for new customers, that they provide labs to test their chillers and obtain verification of their efficiency and that they have the finances for continual research and development. Slowly, we were qualifying new members of the oil-free community, OEMs like ThermalWorks and Kingfit.

It all comes back to the term we keep using throughout these chapters: customer intimacy. Simply put, from our point of view, it has to do with alignment, but for other compressor manufacturers, that's not how they approach selling their products. Their primary concern is annual volume sold, and most of the time, the customer strategies are not embraced.

Turbocor couldn't take on the giants of the HVAC industry and force change to the status quo, not on its own. We needed partners to do so. Prices are important, it's true, but when we check in with our customers, when we travel to their headquarters and meet with their higher level executives face-to-face, price is only a small part of that. We talk about it once a year, and then we never mention it again.

What do we talk about, then? We discuss how we can win business together. We deliberate on strategies for building new products and for which markets we want to develop them. We throw out ideas for different ways of enhancing the connections between our companies so that the oil-free community is working in sync, not at cross purposes.

If you look at the bigger companies that make their own compressors, it's a vertically integrated infrastructure. Like car manufacturers who produce the engine and the chassis and then assemble it all, they build their own compressors and their own chillers, which they then sell together—one company, one product.

But in our case, there are two companies and two products. That important distinction can't be overlooked, no matter how closely our compressors and their chillers are allied. We have to make sure we work in tandem, or else we can't compete against those one-stop, vertically integrated shops.

That alliance has to take place at the topmost levels of our customers' companies. Our technology advancements and product development take place so quickly, spurred on by the innovation that has been Turbocor's hallmark since the beginning, that we don't have time to work our way up through the layers of administration within a potential partner's company.

The typical sales method, as used by traditional sales organizations, is to form relationships with the buyers, the people who will actually be purchasing products. Those buyers in the supply chain are key, but transactional. DTC needed to consider the strategy to a greater degree. We went right to the CEO, CTO, or business owners because we had to understand what their business strategy was and how we could be part of achieving that strategy while simultaneously meeting our own goals.

As this interplay was taking place, and thanks to this customer intimacy, we began to notice the early shift in the market that was moving our customers in a new direction. Data centers, which had been slowly expanding, started to take off. Technology companies came to us with special requirements for cooling, like low lift and high suction temperature, in addition to the already available rapid restarting, excellent features for centrifugal oil-free compressors. Most of our customers start to engage with data center applications in the European, U. S., and Asia-Pacific markets.

We understood that it was a big deal and very different from comfort cooling. Turbocor was about to face an entirely new opportunity to pivot into a burgeoning field that was changing the way the HVAC-R industry did business.

Danfoss Turbocor would no longer be cooling people alone, but also sophisticated microchips and supercomputers in a highly innovative and agile business. Energy efficiency would become more vital than ever before, as top data companies were making commitments to reduce their carbon footprint. That was our dream coming true: a powerful group of end users was driving the total life cycle cost and creating a completely new ecosystem, eager to innovate and redefine the collaboration and speed in the HVAC industry. This had never been seen before, but it was coming with new reliability requirements and we had to be agile and ready for a tremendous opportunity.

Scan the QR code below to learn more and view pictures.

# Part Five—2018 to Present: Big Players on Board

CHAPTER TEN

# The Digital Shift: Cooling the Cloud with Turbocor

Massive data centers consume vast amounts of energy to fuel the computing power necessary to run powerful algorithms needed for the artificial intelligence boom. The bigger these facilities, the more machines they contain and the more heat those machines produce. They must be cooled down.

That's where Danfoss Turbocor comes in.

We began encountering these data center projects around 2013 and 2014. Immediately, we knew business would be conducted differently. Turbocor produces compressors and ships them to our customers, so we know where they're going, but aside from the fact that they're being used in chillers and our knowledge of those chillers from talking with customers, we aren't always told for what new applications those OEMs might be using our devices.

Our first inkling was when McQuay, one of the major OEM players and a subsidiary of Daikin, began asking us about rapid restart functionality. The request was an odd one, to say the least. Chillers need to take time to start up after they're shut off. Typically, this is a five- to ten-minute process, allowing the oil used in the chiller to equalize before the unit is in the right place to start again, and it has not been an issue for comfort cooling.

When we dug into this request, they explained to us this was an important function for data centers. If the chillers serving those centers have to wait that long to restart, then they lose the cooling capabilities needed to keep the computers at the right temperature. They needed those minutes compressed to a shorter time span.

How much shorter? A few seconds.

It became obvious to us that our oil-free magnetic bearing compressors didn't have the same issue. There was simply no oil that we had to wait to settle. So we got in contact with the data center people to better understand what they would need from a product like ours. It turned out they were extremely focused on reducing their carbon footprint and energy efficiency, because of the massive amounts of power needed to run those centers. The fact that we are still depending on fossil fuels for energy generation became a big issue in terms of the data companies' commitments to shareholders to reduce carbon emission. Energy efficiency went to another level, and of course, they wanted any cooling devices to be extremely reliable. You can imagine what the consequences might be if a hyperscale data center were to lose cooling and shut down the supercomputers. All AI functionality would be lost.

Starting in 2014 and through the next four years, we continued to foster closer connections with the owners and operators of these data centers to better understand their needs and to educate them—the pull strategy in action, as we do not sell directly to data centers, but OEMs that manufacture chillers. At the same time, we pivoted our product development to better serve this market with a new level of energy efficiency and new functionalities and products like TGS490, TGS380, and VTX.

Efficiency improvements, coming from the investment to build a strong aerodynamic team back in 2017, became higher priority. The improvements relied on high fidelity computational fluid dynamic simulation tools—using 100 Cores of computing power that evolved rapidly to 500 Cores—combined with better laboratory capabilities. Right before the COVID pandemic hit was when the volume in their building and expansion began to pick up. What started out as 5 percent of our total sales expanded gradually to 50 percent of our products today.

Later on, we were able to start tracking this gain because we watched the uptick in orders for compressors capable of delivering a higher saturated suction temperature (SST). The sole reason to make such a request was to support data center infrastructure, because higher SST means higher indoor temperature. Less of a load is placed on the compressor, and not as much power is consumed.

Word of mouth brought us that corner of the market, too, because as we were researching this growing market, companies like Google, Microsoft, Apple, Meta, and AWS were researching us. They saw OEMs like Daikin McQuay, Smardt, Multistack, Climaveneta, Airdale, Geoclima, BSE, and others promoting high-efficiency chillers using new oil-free compressors, compressors stamped with the Turbocor name and guarantee. It dawned on them that although these OEMs had different approaches, one thing was the same. It was as if they collectively said, "Wait a minute, these guys are all using the same product in their machines. Let's talk to the source of this technology."

You see, the computer data being used to run these centers had quite the skill at creating what are termed ecosystems. They bypassed the normal supply chain and information trail, forming new, direct relationships with suppliers and sub-suppliers. We found this made for a faster, dynamic atmosphere in which all could discuss issues and how things could get done better and in a timely manner. The open dialog was refreshing, despite the fact that it came with stringent, inflexible demands. When they needed something, they needed it right then, and everything had better be done precisely.

Google had been monitoring these operations in the U. S. for years, collecting data from various applications in data centers sorted by different chiller brands to see if this technology was really delivering energy efficiency. Once they examined all the data, they were convinced it would be the best option for their needs. We established a strong connection in the U. S. with the Google chiller group and, later on, with the VP of engineering and data center advanced technology and innovation, Jens Peter Clausen (JP)—a Danish connection, but not part of the founding family—and his team with regular meetings.

Three working groups were established, looking for all Danfoss technologies in data centers. Danfoss Turbocor was leading the discussion on chillers, led by Frank Ford. The goal was to standardize their chiller platform, focusing on lower energy consumption, lower or no water consumption, low GWP refrigerants, and high reliability solutions. Most of these discussions involved OEMs, but they would continue to have talks and consultations with us directly, leading to a standardization of air-cooled chillers that could be applied in any climate conditions without need for water. Those chillers could then be produced on a large scale quickly to meet their exponential growth.

Around the same time, we had a request from Google to audit our quality system in September 2021. We did not sell directly to them, but they wanted to know our strengths and weaknesses. The results ended up better than we thought. They did the same with our OEM partners, some of whom scored a little better than us and some lower. It was a great learning experience to better understand what was important to them and a very different audit compared to what we were used to from the OEMs' traditional audits.

It should come as no surprise that firms like Google, which are vacuums of all manner of data, relied on sheer numbers to reach this conclusion. We had been testing and reaching those numbers for years. This was the ultimate realization of our push and pull strategy, the point at which leading data companies recognized the strategic value of our technology and actively requested Turbocor integration by OEMs. They saw it as a direct enabler of their sustainability goals, delivering superior energy efficiency and maintaining peak performance without degradation in the course of time, with a low carbon footprint, all of which led to the goal of reduced emissions and greener operations.

Most of that growth in data center construction using chillers that incorporated oil-free compressors was led, not surprisingly, by the United States, where the technology DTC pioneered had begun to take hold long before anywhere else. As that was happening, Turbocor continued to gain an increasing amount of attention from bigger global accounts, those giants of the industry.

Why the shift, especially around 2018 and 2019? The smaller entrepreneurs like DTC and regional players (what we call early adopters) began claiming a significant portion of the market share. We were all working toward the same vision of oil-free technology, and an ever-growing number of companies was listening, which meant increased scrambling among other OEMs to catch up.

That scramble is evident in the acceleration of acquisitions and mergers that had been taking place, especially during the second decade of the twenty-first century. Daikin had acquired McQuay in 2006. Then, starting in 2010, Schneider Electric bought Uniflair in Italy; Mitsubishi Electric assumed control of Climaveneta; TICA acquired Smardt; Trane bought Arctic Chiller Group in the U. S.; and BSE was acquired by Vertiv. This impacted how DTC would do business.

"The focus in the beginning was that you had all of these smaller companies, early adopters, that were very good at innovating," Frank Ford said. "They would adopt Turbocor, integrate our compressors, and come up with some really clever systems utilizing them, some interesting stuff, but they were never able to focus on scale and volume. It was a matter of figuring out how to compete against the big guys. Well, then the big guys started buying these companies and integrating them into their portfolios, and suddenly the expectations from them changed a lot."

Viewed from another angle, these acquisitions, in some cases, could divert the focus of companies from the specific things they were good at. Take Trane, for example, which was acquired in the 1990s by American Standard and by Ingersoll-Rand in 2008. This merger led Trane away from innovating in HVAC as much as they had been, and while they remained a powerful heating and cooling OEM—one whose business we pursued for years—the rapid expansion of rivals in China and the accelerated abolition of new technologies, even Turbocor's race to get oil-free compressor technology into chillers throughout the market, caught them by surprise.

This is why we saw Ingersoll-Rand spin Trane off in 2020 to become purely focused on HVAC, and UTC did the same with Carrier in the same year. Now they were independent, publicly traded companies that focused on their core business and could go back to innovation and specializing. You could see it in their recent performance in spite of the fact that the competition was becoming fierce.

Carlos Zamudio, who handled accounts with Trane for many years, believed that it was the complexity and elegance of the Turbocor compressor that gave it a significant leg up in that race. "It's a big electromechanical machine with hundreds of functions, many more than just a regular compressor," he said. "So when you look at the reliability of this unit, you have to take into account the compressor, plus its controller, plus factor in the people. What I love the most is that at the end, the product is in the hands of companies like Trane, which provides the people to get us to that super reliability."

But China was and is where the market continued to expand at a shocking rate. Haier's success at advancing oil-free technology made a tremendous difference. Compressors produced by Turbocor were becoming mainstream in China and expanding to the Asia-Pacific market. China's growth was something along the lines of 40 percent a year, and their country's acceptance of the technology far outpaced that of the U. S. and Europe. It was the same kind of market dominance as we saw in electric car production. Of course, U. S. companies like Tesla are well known, but today China produces a significant number of electric cars compared to any other country. Once they settle on conquering a market, they move full speed ahead with brutal, efficient force. That's just how they do business.

The ongoing Chinese expansion meant Turbocor had to increase its agility with its products. We had only two compressor models when we moved to Tallahassee in the early 2000s, the TT300 and the TT400, with very few voltage ranges. This limited our sales to the United States. Europe used 460 volts, Canada 575, and China 380. When we looked at our portfolio, we had to add the 380V to all models in order to stay competitive in China. In those early days, we were dealing with a lot of issues with voltage variances, issues that we do not see any longer as the Chinese grid became as reliable as Europe's and the United States'.

Europe was leading the industry in passing legislation about the use of refrigerants, such as EU F-Gas, which nations began enforcing in 2015. This would restrict use of substances that were proven to be more environmentally harmful. Even a small 2 to 3 percent of the global warming potential could come from the refrigerant in chillers application. The vast majority of the global warming potential is coming from energy efficiency.

We saw a great opportunity to play in this field of new refrigerants (HFOs), a perfect fit for centrifugal compressors, and as a leader, we wanted to be the first to do so. Our labs in Tallahassee began using these chemicals that were designed to have what was called a "zero or lower global warming potential" (GWP). The R-1234ze in particular was a great fit with centrifugal technology, given its thermodynamics property, affording us another advantage compared to screw compressors. One of our early adopters and a great advocate for energy efficiency and sustainability, Geoclima, rushed to introduce this new refrigerant into their chillers, but unbeknownst to us, many aspects of it needed to be evaluated, including flammability.

That alarmed us, especially when we found out this was

being used in chiller units that were designed for the R134a traditional refrigerant. We agreed it would be great for the environment, but we had to test the chemical to make sure it was safe and the appropriate material compatibility.

After hunting around a bit, we partnered with Honeywell to see if we could test and find a clear way to use this refrigerant without safety issues.

Waiting for the test results was nerve-racking. Eventually, after numerous tries, Honeywell reported back to us that the refrigerant wouldn't propagate flame, so the likelihood of it catching fire in the field was next to zero.

This was a relief because Europe was especially keen on using these kinds of zero global warming potential refrigerants. The positive safety results freed up Turborcor to dive headlong with Geoclima into this field, which further enhanced the oil-free community role as an environmentally friendly solution, with options for much lower global warming impact on top of high efficiency. We had already launched the TGS310 in Europe in 2013 with Geoclima, and we immediately got a lot of others following in the region.

Paolo Ferraris, the Geoclima CEO and owner, was an early adopter and a great entrepreneur who was continuously bumping against the innovation and efficiency boundaries. Paolo sold Geoclima in October 2024 to Munters, a Swedish company that specialized in energy efficiency and climate control solutions, targeting data centers. The TGS310 was about to hit the market, so we did a quick material compatibility adjustment. Basically, this was the TT350 compressor with a lot of opportunities to improve efficiency with and optimize aerodynamics for the new R-1234ze refrigerant. The full aerodynamic optimization of the refrigerant came later with the TGS380 in early 2022.

As 2019 was coming to a close, we were not only optimistic about the future of DTC, but especially excited about the future of the oil-free industry. Daily, the world was increasingly understanding how vital it was to reduce harmful emissions and use efficient means of cooling so that productivity could expand and people's lives could be improved. The consolidation of the OEMs and the projected demand for data centers was a perfect scenario for Danfoss Turbocor to grow exponentially. The excitement of the oil-free community was contagious.

The Danfoss Turbocor leadership team and our sales and marketing group had already been to China once that year for an annual trade show, during which we were able to meet in person with our customers, and as October rolled around, we found ourselves back there for our usual planning meeting to build the budget for the next year and review our five-year plan with our customers. This particular meeting was with our Daikin McQuay representatives. We gathered in a clean, bright conference room, exchanging bows and greetings with our Asian counterparts. These were people with whom we'd spent over a decade forming close relationships, and as the year was ending, we were hoping to refine where we were as partners. What would be our plans for the next five years?

Plans don't always work out the way they were intended. I thought about that later, given that we held our meeting in a city few people outside China had heard of but everyone would know about within a year's time: Wuhan.

Our flight back gave us time to think about what our plans with Daikin McQuay would be leading into the 2020s. It wasn't until a week later, well after our return to Tallahassee, that news stories popped up concerning Wuhan and an unknown disease that was exploding.

Nobody knew what to do. The U. S. government had yet to make public note of cases in America, so we were on our own. A handful of us had gone to China, including myself and Lin Sun.

We decided to quarantine ourselves and report to the team if we experienced any symptoms of illness. We were okay, but the incident taught us a lesson. It was crucial to monitor what was going on and send anyone home who needed to stay away from others. Taking care of everyone's health had to be our top priority.

By January, the COVID-19 virus was bigger news, right when the AHR Expo was to take place in Orlando. None of our people had come down with the virus, thankfully, but our concerns grew. Our people in China were instructed not to come, and we discouraged our customers from attending the event, so as to limit the potential spread of this disease, but a few customers did so anyway. We pressed those attendees to wear masks, keep their distance, and wash their hands regularly in order to protect our employees during the visit to Turbocor.

Meanwhile, we held a meeting in Tallahassee in February 2020 with members of the Danfoss climate solutions division following the AHR expo. It was our chance to show off our solid bottom-up process for five-year planning. And it was necessary because it was clear to the Danfoss Turbocor leadership teams that DTC needed a massive upgrade and increased capacity in the U. S. to support the exponential growth that the OEMs and data center companies were starting to indicate, similar to what we saw in China, with 30 to 40 percent growth, but now globally led by the U. S.

Our factory floor was out of date, especially when compared with the new production lines we'd had installed in our China facility in 2017. Some of the old work stations in Tallahassee were the same ones we had brought from Canada. Where we were producing hundreds, now we were producing thousands per month, and the work stations were literally falling apart. Honestly, we found it a bit embarrassing to show people the factory in Tallahassee, especially the new data center end users, given its age and obsolescence. Shouldn't we have a state-of-the-art space that would awe our customers and be aligned with current ergonomic practices, all while matching the ingenuity of our products? We needed to increase capacity quickly, and the only way to do it was a green field factory.

These were some of the concerns I voiced on the long drive back from Orlando to Tallahasse, and I could tell, based on their skeptical expressions, that we might have an uphill battle on our hands.

The team was very excited after winning another AHR Innovation Award in the Green Building category with the TGS490 compressor, and they did an excellent job of preparing the 2020 to 2025 plan for the Danfoss executives with the following principles:

*Objective:*

Align the Danfoss leaders on the strategic directions of DTC in order to sustain the number one global leading position in oil-free compressors, unleashing the full potential of DTC to continue to transform the air conditioning, heating, and refrigeration industry. The business leaders and front line share a bold mission, focusing on the needs of the customer and being committed to transforming the industry.

It will be executed implementing this bold, rolling five-year plan, focusing on growth, self-financed, and allowing the DTC leadership team to manage the business and not be driven by reactive short-term quarterly / monthly KPIs.

*DTC Recommendations:*

DTC has identified the following three categories and the specific actions needed to adjust our winning journey:

- Growth and customer intimacy
- Agility and innovation to outpace competition
- Run the business like your own.

Treating the DTC business as a truly adjacent growing platform for Danfoss, allowing the DTC leadership team to manage the business as its own business.

The team was enthusiastic after a high-performance year in 2019, despite the early days of the first trade war in 2017, resulting in high tariffs with revenue of $144 million, record earnings, and cash flow. Our plan was to nearly double our business in these five years. This plan segmented the customers into three groups:

- *Group A—Early Adopters:* continue to drive innovation.
- *Group B—Global Accounts:* scale the business.
- *Group C—New Regional OEMs:* new applications, like refrigeration and heat pumps.

The plan indicated that data centers and large OEMs would drastically increase their demand, and our proposal included a brand-new, 145,000-square-foot factory. The meeting was going to take place in our new offices, which had recently opened to accommodate a visit from our customers attending the AHR Show in Orlando. It was intriguing to think back to that 2009 leadership team meeting at Turbocor, when we could barely cram the eager and anxious personnel into a much smaller space.

The meeting was scheduled to last about two hours but ended abruptly not long after it started. "It is not a good time for investments," the president of climate solutions said, earning surprised looks from our team. "We will not invest in the compressor business."

Thirty minutes after we'd begun speaking, our time was up. We were all shocked as we filed out of the room. It was clear each leadership team member needed significant time to boost morale and confidence—Lin Sun, VP of engineering and technology; Sha Ouyang, our financial leader; Jose Alvarez, VP of sales and marketing; Sonjoy Goswami, our VP of operations; and Christina Casey, the HR director.

We were all so confident about our plan and the need for investments to satisfy the growing demand from our customers that we were completely taken aback. We had to believe that it had to do with the exit of the unprofitable and commoditized small compressor business for refrigerators and missed generalization of the oil-free compressor business. Or maybe it was an early fear of the unknown COVID situation. It was hard to believe that we would miss the opportunity we had been working so hard on since the early dreams in Australia or, worse, disappoint our loyal customers who were depending on Danfoss Turbocor. But our attention quickly moved as we knew there was some kind of disease circulating, and rumors of a full-blown pandemic had been underpinning discussions we'd had with colleagues at AHR.

That intuition proved correct in the coming weeks. Cases of COVID-19 exploded in the U. S. and around the world. Hospitals became overrun, and governments saw what was happening. Eventually, the state of Florida called around to various manufacturers, including Danfoss Turbocor, and informed us that we might need to be ready to change what we were manufacturing in the event that we needed to start making respirators.

Respirators? That didn't seem like the purview of a company that built air conditioning compressors, but since we were compressing gasses without oil, it was assumed we could easily make devices that could push air to people who needed to breathe. So, we had to be ready with a small team to develop that kind of medical device in case of an emergency. This switch ended up not being necessary as the government managed to import hundreds of thousands of respirators.

To better prepare our workforce, we ordered thousands of masks in collaboration with Danfoss before things worsened. Our thought at the time was that it would be better to anticipate a worse situation, and as it turned out, that was correct, because there was such a shortage of personal protective equipment (PPE) that we had to share some with the city of Tallahassee. The decision was an obvious one, in any case, because although we wanted to protect our workers, we knew we had to do our part to protect our community, too. That solidarity would be needed in the coming years, more than ever before.

Scan the QR code below to learn more and view pictures.

CHAPTER ELEVEN

# Resilience in Crisis: Danfoss Turbocor During the Pandemic Years

The world was suffering through a pandemic the likes of which had not been seen for a century. Danfoss and Danfoss Turbocor were no exception to the hard times.

We ran the business as if we were under siege. Our leadership team created a virtual war room in Tallahassee led by Christina Casey, our HR director, and VP of operation, Sonjoy Goswami, and a similar setup in China led by Ping Chen, the operation director in Haiyan. The team was tasked with looking at the issues we needed to address every day to protect our team: supply chain interruptions, production slowdown possibilities, and potential illness among our employees. I sat in my office, isolated from everyone else in the facility and from Lin Sun down the hallway, with the heads of the team on the screen in front of me arrayed in squares. Concerned expressions looked back.

Our brand-new offices were empty, going unused. The quiet in the office and the distant rumble of the factory floor were a far cry from the voices of debate raised during our February presentation of the five-year plan and the one hundred customers attending our grand opening of the new offices. We knew we would need to address that future growth, but the immediacy of the pandemic put those plans aside for the moment to take care of our people and customers.

As spring wore on, DTC made drastic changes to extend that protection while still running our operations as best we could. We rearranged some work stations on the main floor. Our employees were separated into shifts, with time in between to sanitize the hands-on equipment. Our team adjusted to this "new normal," to use a popular phrase at the time, with tremendous flexibility and resilience, things we had learned in our journey as a small startup company.

Our supply chain is truly global, and to get parts from China was extremely complex. There were sometimes restrictions by province, so shipments in and out were complicated, but the Danfoss Turbocor team in the U. S. and China was working around the clock to keep production running and serve our customers.

For all our efforts, it began to look like a strong possibility we would have to close our doors. Governments like Canada were shutting down suppliers, such as those who sent us the aluminum blocks needed to build our compressors, or the step-motor valve from India. It was up to us, our dedicated customers, and our parent company in Denmark to write to these governments and suppliers and position ourselves as a necessary industry, not purely for comfort, for key sectors like food storage, the pharmaceutical industry, hospitals, and data centers. We and all HVAC companies were classified in the U. S. as essential businesses, thanks in part to AHRI's lobbying and coordination. We had to keep going, for everyone's sake.

Amid the world's turmoil, Danfoss made a critical decision that would become another major pivot point in Danfoss's history. Danfoss had about forty thousand employees around the world, including in Tallahassee, and about 50 percent worked the production floor. We depended on these employees to keep operations running, and they all depended on us to get their paychecks to support their families. Leadership got in touch with each other and, after evaluating the economic situation, announced that no one would be laid off or furloughed. The shareholders would temporarily defer dividends, and the company leadership also called on all executives to voluntarily cut their salary by 10 percent. Such savings might be enough to prevent any Danfoss employee from losing his or her job due to the pandemic. Eventually, nearly every single executive took the pay cut, hoping it would be temporary but knowing it was the best course for the common good of the Danfoss family. We were all too happy to do so if it meant preserving everyone's livelihoods.

The message to each person, all the way onto the manufacturing floor, was this: *we are going through this difficult time together.* And that message was good not only for our forty thousand employees, but also for the people who depended on them—family members, friends, suppliers, even local businesses where they spent their pay. Estimating that each of the three hundred jobs at Danfoss Turbocor represents ten people down on the line in the oil-free community, between OEMs and suppliers, the decision to keep production running ended up supporting three thousand other people and their families. This is the benefit of being a privately owned family business. We can make decisions based on the long-term benefit and not be driven by the short-term bottom line.

During the pandemic, the Danfoss Turbocor production team carried the weight of the entire business. All the compressors they shipped generated the revenue needed to maintain cash flow and ensure salaries kept getting paid, including for those working remotely. While office staff operated from the safety of their homes, the production team showed up daily, fully equipped with PPE and facing risks to keep operations running. When the pandemic subsided, we took everyone to visit the production floor and personally thank these dedicated individuals for their resilience and commitment—and for keeping the business and the oil-free community alive.

A few months into this orderly chaos, we found our order books had grown 26 percent despite the pandemic. This increase was driven by the digital infrastructure, which was growing faster than predicted on our five-year plan. Speed and agility were of greater importance than ever before. After reflecting on the experience of presenting the five-year plan, including the time it would take to go through normal internal procedures and the possibility it wouldn't be approved again, I decided to call Kim Fausing, who by then was the president and CEO of Danfoss.

We brainstormed solutions and discussed our five-year plan once again, a welcome reprieve from the crisis around us. He agreed to look at the plan, especially regarding the $62 million, 145,000-square-foot green field factory, which we knew had generated enough cashflow to fund with a three-year payback independent of the greater Danfoss ecosystem. Then it was back to the war room as the newscasters spoke sternly about health and safety precautions being undertaken across the country.

A few days later, the quiet work from Kim started to yield results, and the Danfoss financial executives and global service leaders reached out to us for clarification. They peppered us with many thoughtful questions. Toward the end of the call, though, they cleared up outstanding issues regarding the financial elements *and* announced that pre-approval had been given, depending on the final approval from Kim and Jesper (which didn't take long). The new investment plan, including the new production line and the green field manufacturing site, could commence.

All of this was stated in a matter-of-fact tone, but it injected DTC with a jolt of excitement. Our team, led by Sonjoy, immediately started putting together the next steps, such as drawing up detailed plans for this new structure. It turned out to be a prolonged process before we could even begin building. Nothing was easy, during or in the aftermath of the pandemic. Planning and building a new factory and production line, with all the machinery and layout planning that entails, can take two years. We had to find the right designers for such a line, pay suppliers, and hold meeting after meeting both within and outside Danfoss, all complicated by the fact that nobody was meeting in person. Virtual meetings became the norm, which made for slow-going.

Meanwhile, on the production floor, work continued almost entirely unabated. Turbocor was producing approximately 1,300 compressors a month. Managers on the floor were making every effort to keep their workers safe, and it was paying off.

As 2020 ground on and 2021 began, vaccines became available. This posed a complex issue for DTC leadership, given the controversy that was already simmering concerning the vaccines and their development. We soon learned that fewer than 30 percent of the employees working together on the floor were vaccinated. We knew we couldn't demand our employees get vaccinated, because that was a personal decision for each of them, but we were worried about illness spreading throughout our people even in modified conditions.

Education was the key. We brought in people to speak about the vaccine—doctors, healthcare experts, and local leaders. We hoped this campaign would reinforce the importance of protecting each other and everyone's families. The workers proved respectful of each other, and we never experienced a single incident of cross-contamination. We never even needed to stop production during the pandemic.

The situation in our factory in Hayian was not different. Thanks to Ping, Jose, and the Xudong leadership, we kept adjusting and learned how to protect our people while keeping the production running to supply our customers. Once again, the resilience and flexibility of the Danfoss Turbocor team in the U. S. and China were noticed by our customers and the Danfoss organization. During a challenging time, the organization was delivering strong growth in customer intimacy by running the business like our own and with a sense of urgency and agility.

But even as the pandemic began to come under some semblance of control, our production—which had been growing at a fast pace—was hitting a snag. Other industries weren't as agile and customer-focused as ours. One of those was the critical microchip-manufacturing network. They were falling drastically behind, and as a result, supplies lagged for everyone.

The compressors we built relied on several printed circuit boards (PCBs) that were composed of hundreds of electronic components, including the passive ones like resistors and capacitors. In addition, the microchips were the brains the computer boards operated on. Our suppliers began calling us with statements, instead of negotiations. They would allocate us five thousand microchips when we needed twenty thousand, period.

There was no arguing our way to more microchips. The only way to get the amount we needed was on the open market. It turned out that several companies received more microchips than they needed to fulfill their orders, or they were not growing due to the pandemic, so they hosted online auctions where they took advantage of the demand, asking anywhere from three to ten times the normal price.

Seeing that upcharge shocked our budget's bottom line, but the choice was between paying too much for the microchips we needed or not selling as many compressors as we were producing. We went immediately to our customers with a proposal that we hoped would lead to a workaround for the issue: either we could allocate a portion of the compressors they had ordered, short of what they wanted, or they would receive the total order while paying a higher price to compensate our additional expense of acquiring the microchips. It was the cost of doing business during a turbulent time, as industries struggled to recover from the COVID slowdown.

We knew we had to be totally transparent with our customers, because that was how we'd always maintained our close relationships with them. If we wanted that trust to continue, we had to be upfront. The microchip that would have cost us $5 in the past might cost up to $50. There was no avoiding it. We had several special industrial microprocessors per PBC board that added up to our annual volume, an equivalent cost increase of $7 million that we had to share with our customers.

The transparency and sense of urgency ended with every single customer agreeing to pay this surcharge. They understood the situation and didn't want to slow their own projects that required DTC compressors. It was heartening to know that the years of effort we'd put into customer intimacy were paying off in a big way. Even Haier stepped up and found a source of microchips in the appliance industry that we hadn't known about. They bought them for us, allowing us to minimize the pain we all were going through.

Our customers came to our aid even as we were doing our best to keep our promises to them. That solved some of the problem. The rest involved actually getting our hands on the microchips.

It wasn't a simple matter of placing an order from a different shop and awaiting delivery. The companies that had excess chips were running online auctions from all around the world through brokers. The bidding was fierce, and we couldn't afford to take a moment's break, or else we'd lose out. Sharath, our director of supply chain, on his own or through delegating to others, monitored and bid in these auctions 24/7.

At any given moment, one of our people was hunched over their computer screen, watching an auction in Taiwan. They would click on the control to up their bid until they won, but almost immediately another auction listing would appear in Hong Kong. While one of our team members logged in and entered a new bid, notice would arrive of a third auction in Thailand. Hour after hour, day after day, they watched these auctions like hawks, until we had all the microchips we needed and more.

This supply disruption took about eighteen months to destabilize, until we had enough. Eventually, using our experience with how to acquire the microchips, we were able to provide extras for some of the Danfoss divisions in

Denmark that were working in a more conventional way. Once again, our innovative thinking and independence allowed us to be agile in a situation that demanded thinking up unconventional solutions, showing the resilience of a team that had been running the business like their own for several years.

That agility, along with our persistence, paid off as we were finally able to secure delivery to our customers without any interruption. During the pandemic, we kept up regular communication with our key customers, checking to see how their families, employees, and businesses were doing as if they were an extension of our own. Charlie Kenyon, CEO and one of the main shareholders of Multistack, called one day to express to the Danfoss Turbocor team his deep appreciation that we never stopped shipments during this difficult time.

"Danfoss Turbocor is the only supplier that never forced us to slow or stop Multistack production," Charlie said. "You are truly the best partners we could ask for." Multistack had been with Turbocor from the beginning, and it was a great encouragement to our team to hear affirmation like this from a leader like Charlie.

As the pandemic slowed, we were able to secure Trane North America as a customer, a deal almost twenty years in the making, about two years after their integration of Arctic Chiller Group, which they had acquired in September 2019. Our success could be attributed to four key developments: DTC's success with the push and pull global strategy; the exponential growth in China, of which the entire industry was taking notice; Jose La Loggia's belief in Turbocor's oil-free technology, which led Trane Europe to be a successful member of the oil-free community; and the growing pressure of the data center market for better energy efficiency. More companies were embracing magnetic bearing compressors, and as those companies grew, Trane North America was ready to adopt the technology on a substantial scale.

The oil-free revolution was finally mainstream, instead of being on the outside seeking a way in. Further evidence of that was TICA's opening of a brand-new factory for the manufacturing of oil-free chillers in Nanjing in 2022. TICA had started back in 1999 in Nanjing, and unlike many companies, TICA emphasizes institutional and collaborative founding, with strong ties to joint ventures and partnerships such as Carrier, UTC, and Yanmar. Thanks to Jiang Li, a brilliant entrepreneur and businessman, TICA is a global leader and well-respected company.

In the early days, Jiang showed interest in our technology as we were beginning to get our feet under us following the 2009 Great Recession. He came to Tallahassee from time to time back then but decided it was too early to take action. It wasn't until a few years later that he opted to acquire Smardt, headed by Roger Richmond-Smith, and it turned out to be an excellent fit for both entities. Richmond-Smith had been approached by several others but didn't think the culture of those companies matched up with his ideals.

"Good engineering people, good marketing people, good technical and electronics people, good procurement people, good service people," Richmond-Smith said when enumerating TICA's qualities. "And the other thing that I appreciated very much about Jiang Li was his insistence on quality standards and test facilities, which I found very impressive."

The oil-free community was growing by leaps and bounds, and in 2022, we ended the year with $246 million, a growth of 70 percent from the pre-pandemic years. We added another AHR Innovation Award and Product of the Year with VTAC-400, the first hybrid axial / centrifugal compressor in the HVAC industry, which had reduced the size and weight of the compressor by 50 percent.

As the pandemic waned and we continued to get a handle on our supply issues, construction began on the new factory in Tallahassee. Driving by the site on the way to work each morning buoyed our hopes. It symbolized a promising era about to begin, a welcome indication that despite the hardships of the past few years, Danfoss Turbocor was building the future, not merely talking about it.

The future was on my mind, too. What about our Danfoss Turbocor team, our loyal customers, and our entire oil-free community? It must continue its journey, which we had barely begun scaling up. There was still a lot to be done to change the HVAC industry. In reflecting about change, the following became clear: most of our early adopters were acquired by large companies; the leading end user became data centers; the Danfoss family had passed the baton to the third generation; and Danfoss Turbocor had to shift gears from a small insurgent startup to a very profitable business. All of that meant it was time for me to change, too.

As with other leaders that came before me, I served the team and the business, and the mission had to continue. With our children raised, graduated, and pursuing their own flourishing lives, Juliana and I considered that it might be time for us to explore our next chapter, beginning the transition to retirement from Danfoss. But these steps must be taken in harmony with the business, our team, and my family.

I already had definite thoughts about the matter and started to discuss them with Noel Ryan before COVID. Afterward, I continued these conversations with Kristian Strand, current president of Danfoss climate solutions, and shared my thoughts about who the new leader for Danfoss Turbocor should be.

Danfoss encouraged these kinds of considerations as part of our culture. As we learned from the book *Good to Great,* leaders must build to endure, establishing successors who can carry the company forward without disruptions. Even if you're brand new to the company, maybe a younger director or vice president, you're asked to draft a summary of three key people who could replace you if you were to depart tomorrow. They also want to know who would be a good replacement for you in, say, two or even five years. That way, if you were promoted or, heaven forbid, a plane went down and you were on it, the company would know someone skilled could step right into your position.

Eventually, one of three outstanding candidates stood out. Rogerio Federici had been a member of the Danfoss Turbocor team since July 2022. He was part of the leadership team working on updating the five-year plan, as the new vice president of sales and marketing for Danfoss Turbocor. Rogerio had a long career with Danfoss in South America and

Mexico, which culminated in becoming a business leader of our global condensing unit business based in Europe before we hired him in the U. S.

We found him to be ideal because he had strong management abilities and brought a different set of skills. He was not an engineer like most of us on the Danfoss Turbocor leadership team, but an economist, someone able to look at strategic business elements. He had also mastered navigating Danfoss to get things done while preserving some level of independence, even when deviating from the norm. Working side by side made for a smoother transition, of course, as he was able to bounce questions off me in order to gain my advice.

Rogerio's first experience with Turbocor was in 2013, around the same time as the 100-percent acquisition by Danfoss. He was working with a customer who had been buying scroll compressors from Danfoss but oil-free units from DTC as well, which intrigued Rogerio. "I started to get curious," he recalled. "Why are you pushing for this technology? What is the benefit? What is the value proposition behind that? And they were so passionate about it. They said to me, 'This is a unique thing. This is a unicorn.'"

His interest in Turbocor drove him to reach out to us, and having seen how that interest grew over the years, I was reassured by the idea of him taking the helm. The business was growing and remaining financially healthy, with a new five-year road map to get to $500 million revenue. But my greatest concern, first and foremost, was for the people at DTC.

We had a close team, my second family, forged through

years of struggle dating all the way back to the chaotic move from Canada to Florida almost two decades prior. If I were to retire, it would have to be in the right way, setting an example for others who would one day have the honor of "graduating from Danfoss Turbocor," like Roger Federer said in his retirement speech.

The oil-free journey has been marked by countless innovators and pioneers, individuals who played vital roles in developing and promoting this transformative technology. Many have made significant contributions, shaping the path forward. Yet, among them stands one figure whose influence has endured for three decades: Dr. Lin Sun.

Dr. Sun is not simply an engineer; he is a true renaissance mind anchored in strong values, guided by empathy, and driven by an unwavering commitment to balance innovation and execution. It was he who transformed Ron Conry's brilliant concept into a tangible reality decades ago, turning vision into product. When challenges arose, he was the steady presence we sought, offering technical insight *and* human understanding, leading with both intellect *and* heart.

His impact extended far beyond engineering. Dr. Sun cultivated the next generation of innovators, ensuring that the legacy of oil-free technology would not merely survive but thrive. For thirty years, he stood shoulder to shoulder with leadership, influencing strategic decisions that shaped the company's future. His quiet strength and steadfast leadership laid the foundation upon which this technology, and the people behind it, continue to build.

Dr. Lin Sun's contribution is more than technical

mastery; it is a testament to resilience, vision, and the power of empathy in engineering. His legacy will endure, guiding the next chapter of innovation.

I made sure that as I wound down my job and prepared myself for a part-time executive advisor role starting in January 2025, I would be available to Rogerio, Lin, and everyone else for whatever questions they had. I wanted our next actions to be calm and steady, heading into the next phase of Danfoss Turbocor.

There was one big step left to take, perhaps the literal largest one in the company's history.

Scan the QR code below to learn more and view pictures.

CHAPTER TWELVE

# The Danfoss Turbocor Legacy: Our Oil-Free Community

The last few years have been full of milestones at Danfoss Turbocor, seemingly one right after the other, and this progression shows no signs of stopping.

The opening of the Turbocor Configuration and Service Center in Nordborg, Denmark, in November 2022 was a logistical decision but also symbolized a return to the company's roots. Located in the E1 building right behind Danfoss's main headquarters, the site carries deep historical weight. It was there in the 1950s that the very first compressors for refrigerators were produced, thus laying the foundation for Danfoss's rise.

By bringing Turbocor's oil-free, high-efficiency compressors to this historic building, Danfoss united past and future. The move reduced lead times for European customers and enabled local repairs. It also strengthened the company's global footprint, alongside Tallahassee in the United States and Haiyan in China. The company has now established a tri-continental presence that mitigates geopolitical risks and ensures resilience in supply chains.

As Mads Clausen, Danfoss's founder, once said, "If someone is going to make your product obsolete, it's better that it be you." With Turbocor in Nordborg, that philosophy lives on: innovation builds on heritage, customers are served faster, and DTC shapes the future of sustainable cooling.

Back in the U. S., Rogerio Federici was spending countless hours preparing to lead DTC as its new president once I eventually stepped down. We had brought him in from the Danfoss condensing unit business to lead sales in Tallahassee with this objective in mind. Rogerio and I had shared a close professional relationship for thirty years, and I had seen his skills balancing an aggressive business drive with an intense focus on customer intimacy and empathy.

The intimacy Danfoss Turbocor had with customers was the biggest surprise Rogerio encountered when he came aboard. He was used to seeing it between smaller companies and their suppliers, but in our case, every customer received the same treatment. "The customers understand that for them to get the best in terms of developing the future, to be ahead of the competition, they need to be very open to Danfoss Turbocor, because then we can mutually support each other as we grow," Rogerio said.

In the couple of years before I formally retired, I exposed Rogerio to our customer base, making sure he was visible both internally at DTC and externally with the businesses we served. This was a natural extension of Rogerio's consistent presence at these meetings, given his role leading sales. When the announcement came in the fall of 2024 that he would become my successor, there was little surprise.

During this transition process, Danfoss Turbocor hit another milestone: the marking of ten years after the full Danfoss acquisition. Kim Fausing called us with a request to hold a meeting of the entire Danfoss Global Executive Team (GET) in Tallahassee in 2023.

This was an exciting development, in part because it gave us a chance to present the full five-year plan, now updated to 2027 and showing the potential to reach a half-billion dollars in revenue by that time. It was overwhelmingly well received by the GET. After the meeting concluded, we went down to the main production floor for interaction and pictures with the Danfoss Turbocor team, with tours of the new offices and factory under construction.

Watching the GET members and our team interact was a delight. The pride and confidence from our team was evident in the excited way they discussed our production statistics and how well we'd weathered the pandemic crisis. We were able to show them that during the first ten years of full Danfoss ownership, Danfoss Turbocor had tripled its sales volume, quadrupled revenue, and multiplied earnings growth and free cash flow by a factor of twelve.

At that point, 2023 was shaping up to be a record year for sales and earnings, with our cash flow stronger than it had ever been before. By the end of the year, we had sold about 15,000 compressors worldwide: 5,000 in China; 4,500 in Europe; 4,000 in the United States; and 1,500 in the Asia-Pacific market. To put our improvement in perspective, in 2009 DTC sold nearly 3,600 compressors in total. The year 2024 also started with strong demand, and Turbocor added another Innovation Award by the Chinese Refrigeration Show for the extraordinary efficiency of the TGS380. We later estimated that if Turbocor were an independent company and went to an IPO, the cap value to the company would exceed $1.5 billion, a major change compared with the troubled years of the unprofitable joint venture two decades prior. With the recent demand from data centers, we have established a cadence of doubling the business every four to five years. The period of 2025 to 2030 is on a similar trajectory.

Rogerio was already at DTC during that 2023 visit, so it became a well-orchestrated event on our path of transition and a showcase of the growth we were enjoying, the succession plan, and the future of the company. But the real centerpiece of the showcase was the massive new factory that was already underway.

Construction had begun in June 2022, as the world was emerging from the hardest times of the COVID-19 pandemic. Every day, as we arrived on the DTC campus in the Tallahassee Innovation park for work, we could see the progress as heavy machinery converted open ground to a dirt patch; then we watched as a metal frame rose from the site. Eventually, when we all gathered for the grand opening in May 2024, we stood before a massive, sleek white building emblazoned with the bold red word "Danfoss."

Planning for the 2024 event had its own challenges, given the visibility that Danfoss Turbocor had acquired within the

Danfoss group due to its exponential growth. Though everyone was excited to hear about the new facility and eager to join the grand opening event, we were hesitant to invite all who wanted to come from inside the company, for fear of crowding out customers. We had to prioritize the oil-free community and be consistent with our long-term focus on customer intimacy. The list of guests for the new factory opening was focused carefully on the oil-free community, our loyal early adopters, the global accounts, and new customers, along with leaders from our local magnetic ecosystem in Tallahassee and a few key Danfoss executives.

The RSVP confirmations started to arrive in early April, and we saw already that the event would be something very special: the oil-free community, true believers in energy efficiency, and dreamers who saw the potential of Turbocor when few could see it, gathered together and celebrating a collective success, while at the same time showing that this was only the beginning of sustainable transformation.

On the night before the grand opening, we hosted a welcome reception, and the energy in the room was magnetic, literally. The oil-free community is relatively small, but all of us are proud to be part of something greater than ourselves, having changed the HVAC industry forever. At the reception, the vast majority of our customers were represented by their CEOs, the owners and executives of companies that represented over 95 percent of the Danfoss Turbocor revenue from around the world.

The oil-free community was also represented by the Clausen family—Anette, Jørgen, Mads, and Mads-Peter—

along with Danfoss CEO Kim Fausing, Kristian Strand and other Danfoss executives, and of course, our amazing Danfoss Turbocor team. It was a night of celebration and magnetic connections that all enjoyed, amid rising expectations for seeing the new factory the next morning. It would be an opportunity to visualize, literally, the bright future of oil-free technology.

The support from our customers was massive, with participants coming from all around the world: Daikin, Trane, Emcor, Modine, AtmosZero, Mecalor, Thermal Works, Smardt, TICA, Edged Data Centers, CCU, Mitsubishi Electric, Multistack, Quercy Refrigeration, ACS/Mag Zero, AWS, Dunham Bush, ENGIE Refrigeration, Geoclima, ACI, LogBus and HAIER. This turnout was a great boost not only for our team at DTC—especially Mads-Peter, who happily visited with many of his former co-workers and customers—but for the rest of the oil-free community, too. This tremendous honor made it clear to our customers how much the family supported what Danfoss Turbocor had been doing and how the Clausens, Kim Fausing, and the Danfoss board had supported the growth of our operations in Tallahassee, Haiyan, and Nordborg. It also showed the extent to which Danfoss Turbocor represents the Danfoss DNA of innovation, long-term value creation, and engineering a better future.

Mads Clausen, board member and son of Jørgen Clausen, commented that "at every inauguration of Danfoss factories that we have participated in—and there have been several around the world—we never really had such a great experience, a chance to talk and interact with real customers with this degree of intimacy." This was a completely understandable statement. The family and leadership can often be quite removed from the day-to-day operations, given that their attention is focused on running the foundation and managing the board for the good of the entire Danfoss ecosystem.

Altogether, more than one hundred people attended the dedication of our 145,000-square-foot, $62 million state-of-the-art facility, and donned safety goggles so they could take a tour. The modern, energy-efficient building had plenty of natural lighting, reflecting the goal of being carbon neutral by 2030, and was the new home for the TT/TG/TH line of Danfoss Turbocor compressors. Capacity increase, a highly automated production line, and future expansion were built into the design, incorporating everything we had learned from our line built in Haiyan in 2017. But this did not mean that the older building would fall into disuse—far from it.

Our demand for compressors was so high when the new building opened that we had to keep both factories open for the first several months, allowing the new facility to come fully online while maintaining and then increasing production. When we look back at the five-year plan presented to Kim in 2020 during the pandemic, it's clear that our projections were conservative. By the end of 2025, revenue had nearly tripled, from $144 million to $392 million, which was 44 percent higher than what we originally forecasted, or two years ahead of plan. Along with this impressive growth, we've delivered a stronger bottom line, achieving earnings and cash flow well above the plan's expectations and making return on investments sooner than expected.

It wasn't until 2025 that we were able to finalize the transition from the old factory to the new one. In the meantime, plans are underway to give the twenty-year-old building a new life, one that hearkens back to Turbocor's early, if chaotic, days as a place of unbridled experimentation. It will become an expansion of the existing R & D labs and prototype workshop, a place where that kind of imaginative innovation can continue. Engineers there can play with their designs without distracting the main building from high-quality, high-volume production of Turbocor's newest and best compressors.

We think of this as a physical separation of innovation and execution, striking the right balance between the two and maintaining that balance to bring DTC growth. This separation lends a hand in developing new, innovative products in the market at the same time that we scale up our production of our standard product to meet the high demand at our new facility, as is the case with our latest compressor, the TG380.

This compressor is the culmination of decades of design, building on the magnetic-bearing system Lin Sun developed. We optimized that design using many simulation tools, similar to the kinds employed in the testing of aerodynamic elements of jet engines in order to squeeze greater efficiency into a smaller frame. It also incorporates many needs expressed to us by data centers, such as the ability to run at higher temperatures and the use of refrigerants with lower global-warming potential. It's good for the environment and has unbelievable efficiency compared with any other designs in the compressor industry. In fact, the reaction from many of our top OEM customers when they saw the initial reports of the TGS380 was, "Is this right? How could you make something really good even better?" And these comments were coming from key players who produce their own compressors.

Frank Ford, now our head of sales and marketing, is happy to have been part of this expansion that started back during his product management years. "Today we have sixteen different compressor models and twenty-nine different compressor variants," he said. "The variants are those that can operate to meet the specific needs from different verticals like data centers. It's quite a big expansion of our portfolio."

This is far from the end of the story for Danfoss Turbocor. As the HVAC industry continues to realize the vital nature of oil-free technology, we can expand to entire sectors that haven't touched on these advantages yet.

Consider refrigeration. DTC has been so focused on comfort cooling, and recently data centers, that this market has not yet encountered the benefit. Consider medium-temperature fruit storage cold rooms or indoor farming. Fruit harvesters bundle together their apples and store them in giant rooms that make use of a median temperature preservation process to keep them from rotting. These rooms exercise tight control over both temperature and humidity, and sometimes low oxygen levels, with the end result that huge amounts of apples can be stored without degradation for up to eight to ten months. Doing this requires a massive amount of precise cooling.

A few European countries have begun looking at new technologies to support this expenditure of energy with efficient compressors, but our industrial refrigeration industry is so conservative that in many cases, the technology being used is fifty or sixty years old. This is where DTC comes in, with the potential to save 40 to 60 percent in energy consumption in medium-temp refrigeration applications.

"We can take the same technology and deliver incredible efficiency in a lot of different ways," Ford said. "Refrigeration is one of them. Heating another. Even having the ability to start taking this outside of traditional air conditioning and refrigeration lets us see opportunities everywhere. We just have to be smart about which ones we go after."

Alongside this, we have a new product coming that utilizes active cooling: the heat pump application mentioned in previous chapters. The goal is the complete electrification of heating so that all boilers using oil or natural gas as fuel can be replaced. The compressor for this system runs in reverse, allowing the production of heat, which will be revolutionary for many countries around the world that are working hard to reduce their dependence on natural gas and other fossil fuels. This is especially a concern in our European market, given the convergence of multiple factors during the post-pandemic years: structural dependence on imported natural gas, years of underinvestment in supply, tight global LNG markets, constraints in nuclear capacity, the pace of renewable deployment, and ultimately, Russia's invasion of Ukraine, which further intensified these vulnerabilities.

This is where our continued education about oil-free magnetic bearing compressors comes in. As electricity prices continue to rise, and as new sources of power struggle to be built in time enough to satisfy the growing demand, energy efficiency becomes increasingly crucial. We must keep demonstrating that in cases like these large refrigeration systems, we can save between 40 and 60 percent in energy usage and be financially attractive. Safeguarding the environment is an admirable goal. But real change happens when solutions make economic sense. Decarbonization must also reduce costs, strengthen energy security, and improve competitiveness. The return on investment of oil-free compressor technology simply can't be beat."

There's constant discussion at DTC about what is coming next. We fully expect this conversation to be enhanced by the conversion of the venerable older factory into an innovation workshop. Looking back to 2009, when the company trimmed back the overwhelming numbers of projects underway and focused on a few, it should come as no surprise that we never got rid of any of those designs. We're still contemplating some concepts that force us to ask ourselves key questions: Do we want to go with bigger compressors? Do we want to go smaller? What are the applications those would be used for? Other than the HVAC industry, in which we're already established, what other industries could our technology be useful for, that nobody has thought of yet?

Since our foundational patent we vigorously defended has now expired, it has fulfilled its critical role of protecting our innovations during the early stages, allowing us to validate the concept, mature the technology, and scale the business. We continue to produce patents and have several strong new ones. However, while patents are important, our tightly-guarded trade secrets are equally vital. Unlike patents, which disclose methods and invite workarounds, trade secrets protected by robust NDAs and non-compete agreements safeguard our proprietary manufacturing processes, material specifications, and most critically, the source code that powers our advanced mechatronic algorithms.

Another key differentiator is our commitment to vertical integration. By designing or producing all critical components in-house—including magnetic bearings, sensors, shafts, motors, electronics, power systems, and aerodynamic elements—we maintain superior control of quality, cost, and reliability. This integration enhances performance and fortifies our competitive edge.

Naturally, our success has attracted attention. Some companies have been pursuing similar technologies for over two decades, while in China an emergent effort is attempting to follow our approach in the same capacity range. Meanwhile, traditional OEMs remain focused on high-capacity centrifugal systems in the 500-ton-plus range, typically limited to low-pressure ratio or water-cooled applications, that are produced in relatively small volumes.

We welcome competition. Expanding the oil-free, high-efficiency market benefits all stakeholders, especially the environment. As awareness grows, so does demand, and our unmatched experience and broad portfolio position us to lead. Competition drives us to innovate faster, deliver better quality, and remain cost-competitive, ultimately making us a stronger partner to our loyal customers.

Importantly, the rise of Turbocor's high-efficiency oil-free technology has also raised the bar for traditional lubricated systems. Technologies like screw and scroll compressors are now being pushed to improve their own energy performance in order to remain relevant. In effect, Turbocor has redefined the industry benchmark, forcing all technologies to operate with less energy. This shift is also about setting a new standard for sustainability and performance that the entire HVAC industry must now strive to meet, with tremendous benefit to the environment.

Equally important is our customer intimacy. We stay close to our partners, understand their evolving needs, and co-innovate to solve real-world challenges. This collaborative approach strengthens relationships and also fuels meaningful innovation that keeps us ahead of the curve.

The Turbocor revolution is in full motion. What began with a few visionaries in modest workshops has evolved into a global movement. Pioneers like Ron Conry, Lin Sun, York Zha, Charlie Kenyon, Carlo Grossi, Ms. Wang, Ron Hallon, Roger Richmond-Smith, and Jørgen Clausen have played a key role in convincing the industry that high efficiency and environmental responsibility can go hand in hand.

Yes, we have more competitors now, and that's a good thing. Every new entrant advocating for oil-free technology grows the market and reinforces our leadership. We must not fall into the trap of complacency. We thrive on challenge; it keeps us sharp, agile, and relentlessly focused on testing the boundaries of what's possible.

For Turbocor, every hour of operation has been a quiet demonstration of circular economy principles in action: less energy drawn from the grid, fewer emissions released upstream, and a lifetime of efficiency that compounds year after year. As the technology matured, another dimension of circularity revealed itself: the compressors simply didn't age the way traditional machines did. With no metal-to-metal contact, wear was minimal, and performance stayed remarkably stable over time. This longevity wasn't accidental; it was engineered into the DNA of the product. Lightweight components, precision manufacturing, and the increasing use of low-carbon "green aluminum" all contributed to a design philosophy that valued durability over disposability.

And now, as production expands globally, that philosophy extends to the factories themselves. After proving the model in Denmark—where Turbocor compressors are already produced in a carbon-neutral facility—Danfoss is configuring new manufacturing lines in the United States and China to operate with the same carbon-neutral footprint by 2030. The circular economy is no longer just embedded in the product; it is embedded in the way the product is made.

A further turning point in the Turbocor journey came with the transition to low-global-warming-potential refrigerants. By enabling systems to operate with refrigerants that carry dramatically lower or no climate impact, Turbocor technology helps eliminate the environmental risks associated with potential leaks—an issue that has long challenged the HVAC industry. This capability has become one of the primary reasons for the rapid global adoption of Turbocor compressors. Customers, regulators, and system designers increasingly seek solutions that not only perform efficiently but also align with tightening environmental standards.

But perhaps the most powerful expression of circularity appears at the end of each compressor's first life. Instead of being discarded, Turbocor units return home. They are disassembled, inspected, renewed, and reborn through a rigorous remanufacturing process that restores them to original performance. High-value materials are preserved, waste is minimized, and the compressor begins a second life with the same reliability as the first. This closed-loop model—designed from the outset, not added as an afterthought—embodies the essence of the circular economy. It shows that industrial equipment can circulate through multiple lifecycles, delivering value again and again while consuming only a fraction of the resources. In many ways, the remanufacturing line is where the Turbocor journey becomes more than a story of innovation; it becomes a blueprint and reference for circular economy.

Ultimately, every new oil-free compressor deployed contributes to a cleaner, efficient world. From data centers to hospitals and airports, the ripple effect of innovation spreads—accelerating progress, reducing emissions, and transforming industries.

The Danfoss Turbocor journey has changed the industry and inspired Danfoss itself. As the Danfoss Business System showed Turbocor how to have discipline and execution while reining in the distractions that come with producing an excellent product, so too has Turbocor's wide-eyed enthusiasm for experimentation shown Danfoss the importance of leaving room to innovate with customers and act as an entrepreneur in a large organization. Under the new Danfoss operational model, a level of freedom is given to local business leaders in various branches of the company. This allows for greater flexibility in working closely with customers and taking calculated risks, with less heavy corporate involvement, so that the local leaders can run the business like it's their own.

We had a visit from a retired Sabroe compressor engineer, to the Tallahassee facility in early 2025. He was excited by the technological advances and intrigued to see how manufacturing processes had changed throughout the decades. As he was leaving, he told us, "This is something I have never seen in my fifty years of experience. You have no idea what Danfoss Turbocor has done for this industry, but you're going to see it in the future."

The future of oil-free technology in the HVAC world and beyond belongs to those who are willing to rely equally on tradition and vision, balancing innovation and execution. It belongs to the entire oil-free community that Danfoss Turbocor Team and Danfoss have had the privilege to embody in our culture of engineering a better future.

Scan the QR code below to learn more and view pictures.

# Acknowledgments

I want to express my deepest gratitude to my family. To my wife, Juliana, whose unwavering support carried us from Porto Alegre to Monterrey, Baltimore, and Tallahassee: Thank you for walking every step of this journey with me. To my children, Pedro, Matheus, Daniela, and my grandson Helio: You gave purpose and joy to all the long years of work. To my parents, Enio and Lya: Thank you for the education and values that shaped who I am.

My sincere appreciation goes to Danfoss and the Clausen family for the trust and the opportunity to lead the Turbocor team for so many years as if it were our own business.

To our oil-free community and to our customers around the world: Thank you for trusting us, dreaming with us, taking risks with us, and innovating alongside us. Without your belief and partnership, Turbocor would not be in business today.

And finally, to the incredible Danfoss Turbocor team, including every single person who put their talent, energy, and dreams into this purpose: You transformed a bold idea into a global movement. This book belongs to you as much as it does to me.

# About the Author

Ricardo Schneider is a husband, father, grandfather, engineer, and business leader whose career has been devoted to advancing a more efficient and sustainable HVAC-R industry.

Born and raised in Brazil and shaped by a global career, he spent thirty-four years with Danfoss—more than half of them leading Danfoss Turbocor—where he helped turn oil-free magnetic bearing technology into a worldwide standard for cleaner cooling and heating.

An engineering graduate of the Universidade Federal do Rio Grande do Sul, with executive education from Harvard

Business School, Ricardo built his leadership on collaboration, collective intelligence, and the courage to challenge convention. These values guided him as he scaled breakthrough technologies, united teams across cultures, and nurtured the global oil-free community of believers, engineers, customers, and partners committed to innovation and sustainability.

He believes deeply that people and technology can improve the quality of life without harming the planet, and that true progress is measured not only in products or financial results, but in the determination, courage, and collaboration of the people behind them. As he often says, "Great teams deliver great results."

Ricardo and his wife, Juliana, live in Tallahassee, Florida, where they devote their time to family, education, the natural world, and work that leaves a legacy of stewardship for future generations.

Renown

PUBLISHING

Renown Publishing is a storytelling atelier for leaders and families whose wisdom is meant to endure.

Most people sense they carry something worth passing on. What they can't see is the pattern running through it. The meaning they've been living but never named.

Renown's eighteen-month process helps authors discover that meaning, then shape it into a book that does what instruction alone cannot: it transforms readers by showing them someone who walked the road first. Story, not advice, is how wisdom actually transfers.

Working with a small circle of authors each year, Renown's team of award-winning writers and editors approaches each project as an act of stewardship. The result is a book. But the deeper result is an author who understands their own life differently and knows exactly what they're handing down.

These books are not for everyone. They're for the few who will be marked by them. Who will carry what they've read into the rest of their lives. Lasting movements don't begin with crowds. They begin with the people who are never the same afterward.

To learn more, visit renownpublishing.com.

Made in the USA
Las Vegas, NV
27 July 2026

51789878R00135